Science Teachers' Innovative Work Behavior: Factors and Actors

# ERZIEHUNGSKONZEPTIONEN UND PRAXIS
# EDUCATIONAL CONCEPTS AND PRACTICE

Edited by Gerd-Bodo von Carlsburg

VOLUME 89

Palmira Pečiuliauskienė / Lina Kaminskienė

# Science Teachers' Innovative Work Behavior: Factors and Actors

**Bibliographic Information published by the Deutsche Nationalbibliothek**
The Deutsche Nationalbibliothek lists this publication in the Deutsche Nationalbibliografie; detailed bibliographic data is available online at http://dnb.d-nb.de.

**Library of Congress Cataloging-in-Publication Data**
A CIP catalog record for this book has been applied for at the Library of Congress.

The publication of this monograph was funded by Vytautas Magnus University.

Reviewers :
Dr. Martina Möller, Professor at the University of Gießen
Dr. Andris Broks, Professor Emeritus of the University of Latvia
Dr. Konstantinos D. Chatzidimou, Assist. Professor at the Aristotle University of Thessaloniki

ISSN 0723-7464
ISBN 978-3-631-85708-3 (Print)
E-ISBN 978-3-631-87335-9 (E-PDF)
E-ISBN 978-3-631-87336-6 (EPUB)
DOI 10.3726/b19459

© Peter Lang GmbH
Internationaler Verlag der Wissenschaften
Berlin 2022
All rights reserved.

Peter Lang – Berlin · Bern · Bruxelles  New York · Oxford · Warszawa · Wien

All parts of this publication are protected by copyright. Any utilisation outside the strict limits of the copyright law, without the permission of the publisher, is forbidden and liable to prosecution. This applies in particular to reproductions, translations, microfilming, and storage and processing in electronic retrieval systems.

This publication has been peer reviewed.

www.peterlang.com

# Contents

Acknowledgment .................................................................................... 7

Introduction ............................................................................................. 9

1. Theoretical background of the innovative work behavior of
   science teachers ............................................................................... 15
   1.1. Innovations in science education: The issue of Diffusion theory ....... 15
        1.1.1. The concept of innovation ................................................ 17
        1.1.2. Theories of innovations ................................................... 20
   1.2. The concept of innovative work behavior ............................................ 23
   1.3. Psychological background of innovative work behavior ..................... 25

2. Methodology ..................................................................................... 33
   2.1. Basic paradigms of science teachers' innovative work behavior ........ 33
   2.2. Research methods and the design of research articles ....................... 34

3. Empirical insight of factors and actors of science teachers'
   innovative work behavior ............................................................... 37
   3.1. The internal structure of science teachers' innovative work behavior  37
   3.2. Comparative analysis of demographic and educational factors
        on science teachers' trying out and sharing new ideas ...................... 51
   3.3. The role of organizational commitment in the innovative work
        behavior of science teachers ................................................................. 65
   3.4. The role of self-confidence in teaching science in the innovative
        work behavior of science teachers ........................................................ 76

3.5. The influence of organization leadership support on science teachers' innovative work behavior ....................................................... 92

3.6. The associations between professional development, professional, demographic factors and the innovative work behavior of science teachers ................................................................ 107

## 4. Discussion ........................................................................................... 153

## 5. Conclusions ....................................................................................... 163

Approval ...................................................................................................... 167

List of figures ............................................................................................. 169

List of tables .............................................................................................. 171

Bibliography .............................................................................................. 177

# Acknowledgment

We would like to thank everyone who has supported and helped us with completing this monograph. Your assistance has been invaluable to us during this process. We also greatly appreciate the financial support of Vytautas Magnus University for preparing and publishing this monograph.

# Introduction

Educational innovation plays an important role in society. As stated by Serdyukov, "For an individual, a nation, and humankind to survive and progress, innovation and evolution are essential. Innovations in education are of particular importance because education plays a crucial role in creating a sustainable future" (2017, p. 5). Fulfilling educational innovation requires understanding the meaning of innovation. According to Brewer and Tierney (2012), innovation has two components: new ideas and the changes which results from the adoption of new ideas. "Creativity is thinking up new things. Innovation is doing new things" (Levitt, 2002). Amabile et al. (1996, p. 1154) state that "all innovation begins with creative ideas." The ability to generate new ideas is not enough for their successful implementation or to put innovation in practice. People who are full of new ideas often do not understand how to implement them in practice (Levitt, 2002).

The implementation of innovations in practice depends on the persons' work behavior, especially how innovative their approach to work is. Scholars state that innovation behavior is a crucial element of organizational innovative development—the best way to foster innovation for success (Chang, 2018, p. 18; Aziah & Al Amin, 2018, p. 426). The permanent relationship between innovation and science education and its function has been under observation over a long time. Layton (1986) states that innovation has become a permanent feature of science education not only in curriculum content but also in the associated teaching methods and materials. "Innovation in science education is less a characteristic of a particular period in time than normal and continuing process. The rapid advance of scientific knowledge and the emergence of significant technologies alone require that this is so […] Even so, the events of the past thirty years provide examples of planned innovation on a scale rarely witnessed previously" (Layton, 1986, p. 9).

For this reason, researchers address the phenomenon of innovative work behavior (IWB), which involves creating and incorporating something new into existing work (Aziah & Al Amin, 2018; Chang, 2018; Pudjiarti & Hutomo, 2020). Innovative work behavior manifests itself in a variety of activities, such as generating, promoting, and realizing new ideas (Aziah & Al Amin, 2018; Sun & Huang, 2019; West & Farr,1990). Innovative work activity requires the ability to think and do things differently while implementing innovations (Pudjiarti, & Hutomo, 2020). The past three decades have demonstrated growing calls for innovations associated with science teaching methods. In light of science education reforms (NRC, 1996, 2000, 2012), teaching methods have focused on innovations that have alternately been called scientific inquiry, discovery, and constructivist approaches (Furtak & Kunter, 2012). How these challenges are reflected in educational practice is revealed through systemic measurement of innovations: the New Consortium Media (Adams et al., 2018), Measuring Innovation in Education Monitoring (OECD, 2019), An Innovation Survey (Halász, 2018), and Olso Manual (2018). OECD (2019) uses The Trends in International Mathematics and Science Study (TIMSS 2015) data for the secondary analysis of educational innovations in science education and presents the results of a longitudinal study about the implementation of innovations.

Researches on innovations in science education discus about the application of new technology (Arici, Yildirim, Caliklar, & Yilmaz, 2019; Liu et al., 2017; Osunkwo & Enyaosah, 2016), the need to infuse arts and social emotional learning content into science education (Bardone et al., 2017), the need to infuse social emotional principles (Garner et al., 2018), promotion of social creativity toward novel student-based solutions and innovations in science education (Aksela, 2019), and the need to use innovative models to improve critical thinking skills and self-efficacy of preservice chemistry teachers (Rusmansyah, Isnawati, & Prahani, 2019). However, there is a lack of research on innovative work behavior of science teachers' and personal, cognitive, and environmental factors. In summary, the lack of research on science teachers' innovative work behavior and its determinants and especially on individual and environmental determinants that led to the problem focused on in the present study: What are the associations between science teachers' innovative work behavior and personal, cognitive, and environmental factors.

The objective of the present research is to study the innovative work behavior of science teachers.

The aim of the present research is to reveal the associations between science teachers' innovative work behavior and personal, cognitive, and environmental factors.

Objectives of the present research are as follows:

1. To reveal the structure of science teachers' innovative work behavior according to Rogers' Diffusion theory.
2. To disclose the role of demographic factors on the innovative work behavior of science teachers.
3. To highlight the influence of organizational affective commitment on innovative behavior of science teachers.

4. To reveal the influence of science teachers' self-confidence in teaching science on innovative behavior.
5. To disclose the influence of organizational leadership support on science teachers' innovative work behavior.
6. To highlight the role of professional development content, duration, gender, and teaching experience in teachers' innovative work behavior.

## Structure of this monograph

The aforementioned research objectives have determined the content and structure of this monograph. In the first chapter of monograph, The Theoretical Background of Innovative Work Behavior of Science Teachers, the phenomenon of innovative work behavior is analyzed from a theoretical aspect. This chapter focuses on one theory of innovations – Rogers' Diffusion theory and three psychological theories: social learning theory, social cognitive theory, and Expectancy-Value theory.

In the first paragraph of *Innovations in Science Education: The Issue of Diffusion Theory*, we discuss Rogers' diffusion theory. Diffusion is essentially a social process through which people talking to people spread an innovation (Rogers, 1995). According to Diffusion theory, the decision process is the mental process which consists of five stages: knowledge, persuasion, decision, implementation, and confirmation. At different stages, an innovative work behavior is reflected by a series of activities in which individuals generate novel ideas, solve practical problems at work, and achieve positive effects.

We analyzed these activities in the second paragraph, *The Concept of Innovative Work Behavior*. In the last paragraph of the chapter, *Psychological Background of Innovative Work Behavior*, we give a psychological background of this concept. In this study, we relied on Social Cognitive theory and examined the role of organizational commitment and organizational leadership in the innovative behavior of science teachers. In addition, we analyzed the links between personal cognitive factors and innovative behavior of science teachers. In this study, we also relied on Social Learning theory, in analyzing the role of personal sociodemographic factors in science teachers' innovative work behavior. Expectancy-value theory was also critical, helping us to reveal the essence of science teachers' self-confidence and to discover the associations between science teachers' self-confidence and innovative behavior.

In the second chapter of the present monograph, *Methodology* we introduce the philosophical and empirical commitments of this research.

In the third chapter of the monograph, *Empirical Insight of Factors and Actors of Science Teachers Innovative Work Behavior*, we describe the results of factors analysis and ordinal logistic regression. The structure of this chapter corresponds to the research objectives.

In the first paragraph, *The Internal Structure of Science Teachers' Innovative Work Behavior*, we analyzed the internal structure of science teachers' innovative work behavior using a Confirmatory factor analysis (CFA) and path analysis. Taking the TIMSS 2015 data for Lithuania, we created a hierarchical linear modeling according to diffusion theory: (1) knowledge, (2) persuasion, (3) decision, (4) implementation, and (5) confirmation (Rogers, 2003).

In the second paragraph of empirical part of monograph, *The Comparative Analysis of Demographic and Educational Factors on Science Teachers' Trying Out and Sharing New Ideas*, we analyzed two innovative activities: trying out new ideas and sharing of new ideas. We performed an ordinal logistic regression analysis of the TIMSS 2015 data for Lithuania. We tested the role of three factors—gender, level of education, and the age of science teacher—on science teachers' willingness to try out and to share new ideas.

In the third paragraph, *The Role of Organizational Commitment on Innovative Work Behavior of Science Teachers*, we analyzed the role of environmental factors in innovative work behavior of science teachers of Lithuania. The result of CFA and structural equation modeling (SEM) analysis disclosed different influences of organizational/affective commitment on science teachers' innovative behavior activities. We detected that organizational/affective commitment had the biggest influence on science teachers' activity of sharing of innovative ideas and the lowest influence on the application of new ideas in education.

In the section, *The Role of Self-confidence in Teaching Science on Innovative Work Behavior of Science Teachers*, we analyzed the role of personal and cognitive factors in the innovative work behavior of science teachers of Lithuania. Our SEM results confirmed the influence of science teachers' self-confidence in teaching science on their innovative behavior activities. The self-confidence of science teachers in teaching science has a greater influence in their applying activities but a smaller influence on the activities of generating and sharing ideas.

In *The Influence of Organization Leadership Support on Science Teachers' Innovative Work Behavior*, we analyzed the role of environmental factor—organizational leadership – using confirmatory factor analysis and SEM. The results of our study correspond to social cognitive theory.

In the last paragraph of the empirical section—*The Associations between Professional Development, Professional, Demographic Factors' and Innovative Work Behavior of Science Teachers*—we performed an in-depth analysis of role of professional development in science teachers' innovative work behavior. We analyzed the role of these factors: DPD—duration of professional development; $CPD_i$—content of professional development; GEN—gender of the teacher; PE—professional experience (number of years the teacher has been working). We also performed a comparative analysis of these factors on the basis of TIMSS 2015 data for different countries: Lithuania, Singapore, and Sweden.

In the *Discussion* section we present the main results of the various sections of the monograph. Our discussion involves a summary of the main findings of the monograph, followed by our interpretation of these results in light of our literature review presented in the introduction and first chapter of monograph. In the discussion section, we examine the results, and draw inferences and conclusions from them.

# 1. Theoretical background of innovative work behavior of science teachers

## 1.1. Innovations in science education: the issue of Diffusion theory

*The section focuses on Rogers' Diffusion theory of innovations, which treat innovation as the process of five stages: knowledge, persuasion, decision, implementation, and confirmation. The section explores the concept of innovations and analyze the challenges involved in developing and implementing innovations.*

Educational institutions are encouraged to develop and sustain dynamic capabilities (Jurksiene, & Pundziene, 2016; Kareem & Alameer, 2019; Fenech et al., 2021) in order for them to respond to the changes taking place in the modern world. In the field of application of educational innovations, the problems and challenges that educational institutions face in the implementation, use, or development of innovative technologies are a routine phenomenon. In search of solutions to these problematic situations, it is necessary to study the experience of educational institutions in the development, implementation, and application of innovations (Sahin, 2006). However, the application of innovative ideas can be useless and even harmful if there is no clarity on the purpose or rationale for implementing those innovative ideas. For the education system, it is important that innovations are scaled (Shelton, 2011). However, one of the limitations of the scale is related to insufficient understanding of the nature of the innovation process as a phenomenon and too few insights into how innovations are created, implemented, and disseminated in an organization. This is one of the reasons why we focus on Rogers' theory of diffusion of innovations (2002, 2003).

For the spread of innovations, collaboration and society support are crucial (Serdyukov, 2017). Uncertainty related to innovations is usually very high (Rogers, 2003), and schools are inclined to routine and tradition. Educational systems and particularly secondary schools remain more conservative than higher education institutions (Gibbons & Silva, 2011) devoting more to students' well-being and safety than to their preparation for real life and work.

More productive and more structured collaborations, such as networks and various consortia and agreements between different research groups, institutions, and organizations, are the expression of changes initiated in current innovation systems (Hekkert et al., 2007; Russell & Smorodinskaya, 2018). The concept of innovation networks can be understood as the internal structure of a self-organizing system, which consists of several parts: the relationship or interface between those parts, and the way or model of operation which is a result of this relationship or interface. The main element of the network is the human being. For this reason, all networks are regarded as social and thus reflecting a unique structure of interrelationships, activities, and practices. Innovation networks have all the characteristics of social networks; hence, the result of social interaction is relevant in innovation networks – innovations, specific actors of innovation networks, exchange of information, and strategies for cooperation activities that promote innovation.

The creation and application of innovation is not merely a field of interest for researchers or an industry or a business. This field is multilayered and diverse, populated by members from different types of institutions: the state, public authorities, business partners, research institutions, innovation service organizations, agencies, economic development agencies, and business associations. The view is emerging that innovation is a process of progress, during which companies or institutions and their partners cooperate in a purposeful manner. When discussing specific innovations, it is usually assumed that innovations are only about creating new products, services, or processes. This is a misconception and projects the concept of innovation as being too narrow in scope. Most innovations are well-tailored discoveries that lead to the creation of new products or services, and, in the process of products, products are improved or adapted to meet other emerging needs of customers and, thus, providing improved versions of products to expand existing market reach or create new markets.

Innovation has always been important because it helps to introduce positive changes into how systems and organizations function. Educational institutions are encouraged to develop their capacities to respond to the changes taking place in the modern world. That is not to say they have to become extremely dynamic and flexible to cope with unprecedented transformation, as illustrated by the pandemic situation and remote learning. However, the problematic situation of educational institutions is related to the continuous changes in the educational sector, which creates a feeling that the development, implementation, and application of innovations is unstoppable. However, without delving into the purpose of innovations being implemented, their application can be useless and sometimes harmful, such as the slow and/or ineffective development and application of innovations in educational institutions, the unfavorable attitude of educational community toward innovation in education, and an attachment to traditional solutions and practices.

Applying innovation to the educational system is a serious challenge. It is not easy to know when innovations can be considered ready for deployment and determine the timeframe within which their impacts are to be assessed. Despite these challenges, innovations in education keep emerging and are developed and applied all the time, although not on a scale significant and visible in the education system and the changes or impact they bring about on everyday activities are rather small and insignificant. There is lack of knowledge on how innovation is created and how it can be used in education. There is no one right solution for implementing an idea or assessing the usefulness of its impact. It is a process encompassing several stages—from the stage of generating a new idea to its successful implementation and later to post-implementation evaluation. Put in other words, the process involves innovation creation, implementation, and dissemination, in direct relation to the maturity and favorable environment of the education system, and the identification of the roles of educational institutions, exsting support for innovations, and the fragmentation of innovations. The successful educational activities of an educational institution and innovative features of culture formation are closely related to the head of the educational institution, their approach to innovation, their understanding of staff needs, and their conviction that innovation plays an integral part in the educational institution's quest for providing quality teaching. In an educational institution, one of the main activities of the teaching staff is to create, develop, and sustain an innovative work culture within the educational institution. There is a growing perception that innovation is a process of progress in which organizations and their partners work together in a target- or goal-oriented manner. Innovation is the engine of progress. Innovation is crucial for the development and success of any organization, and the education system is no exception. Innovation helps to bring about change, as it is often a catalyst for change. Creating innovation is a continuous activity, so any organization willing to change should form a culture open to innovation, involving the whole community, in the search for and development of new and original solutions.

### 1.1.1. The concept of innovation

In order to understand the progression of the idea of *innovation*, it is helpful to review its etymology—the term *innovatio* comes from Latin, which means *renewal*, and dates back to the 13th century (Godin, 2008). In the middle of the 15th century, the word "*inovacyon*" was used in France, which means updating or giving a new image to an existing object.

The scientific literature provides a variety of definitions of innovation. At the beginning of the 20th century, the Austrian economist Schumpeter (1963) introduced the term *innovation*. Schumpeter highlighted the five most important indicators defining innovation: a new or improved product, a new production method, a new market, new sources of production processes, and a new way of organizing activities. Schumpeter stressed that innovation is more of an economic phenomenon than it is technological. Whatever technological invention or discovery takes place in the production process, it can never be considered an innovation unless it leads to or affects economic growth or increases net profit. In order for a company to receive a net profit from innovation, it is necessary that innovation create and disseminate unique advantages compared to that of the participants of internal or external (international) markets. Schumpeter began to see innovation as a change whose main goal is the consumption of new forms of consumer services, goods, production and vehicles, markets, and industrial and business enterprises.

According to Drucker (1985), innovation occurs not only because of ideas that have arisen but also because of the work done in a planned, methodical, and rational manner. Innovation is the ability to see change and to apply that change in practice. He regards innovation as a management tool that uses change as an opportunity to create new businesses, products, and services that generate significantly higher profits.

Following Jakubavičius et al. (2003), the word *innovation* should be understood as a never-ending process and innovation the end result of that process. Here "new" is used to denote something that was used for the first time and was created and used or resulted an expression of human activity that met the needs of society. Depending on the nature of the needs, innovation can mean new products; new technology' a new economic, social, organizational, management structure; and, to sum up, influence the totality of human interactions in work activities. According to Valentinavičius (2006), the concept of innovation is related to the activity. The activity is the result of the application of scientific inventions. The concept of innovation has changed over time—in the middle of the 20th century, innovation was only considered to be the result of the successful work of scientists; yet these days innovation is identified as a process that solves problems, a process that involves coded and unspoken changes in knowledge.

O'Sullivan and Dooley (2009) argue that innovation can be incremental, insignificant, or substantial and radical, and it can only be related to process improvement, which creates added value and contributes to the growth of an organization's knowledge. This view extends the understanding that innovation is an invention, the creation of an original product, or a fundamental change. O'Sullivan and Dooley's statement can thus be interpreted to mean that innovation is also a minor change that takes place in organizations.

Any given definition of innovation introduces new features of innovation, proposes new meanings, and adds what has not been discussed before. Importantly, innovation is always an important phenomenon that has a changing nature, destroying stagnant norms and outdated traditions in organizations.

Cohen and Ball (2007, p. 19) define innovation as a "departure from current practice—deliberate or not, originating in or outside of practice, which is novel." Towndrow et al. (2010, 427) note that "innovations can include changes in policy goals, curriculum design and implementation, assessment regimes, administrative arrangements, leadership, classroom practices, pedagogical technologies and resources, and teacher capacities."

Serdyukov (2017) refers to innovation as sustainable change and admits, therefore, that it is to be regarded as an instrument of necessary and positive change. Any human activity (e.g., industrial, business, or educational) needs constant innovation for it to remain sustainable (Serdyukov, 2017, p. 5). The creation and development of innovation is related to the different behaviors of innovation actors. Social and psychological phenomena can and are already addressed in product, process, service, administrative, social, and technological innovation; radical and growing innovation; incremental innovation; open and closed innovation; and emerging and declining innovation (Rogers, 2003).

Educational innovations are often defined as social innovations. Social innovations are related to the processes of implementing and diffusing new social concepts across different sectors of society (Kolleck, 2014). The term *social* here is important, as it implies the interaction between actors. Thus, social innovations address social problems and challenges and try to offer new solutions to solve them. Educational innovations are social innovations in educational contexts, such as new forms of educational cooperation or novel learning concepts (Kolleck, 2014).

Educational innovations imply nontraditional solutions that are adapted in the existing system, modernizing it and introducing changes. These are types of innovations that have a positive impact on society and engage people in the creative process. In the field of education, analyzing the phenomenon of innovative activity and linking it with teacher activity, Janiūnaitė (2004) defines innovations as a purposeful process of creating, implementing, and using educational innovations at the level of various organizations in order to improve educational systems and form an innovative culture at schools. This approach emphasizes the nature of innovation and highlights the processes of development, implementation, use, and improvement or enhancement of curriculum or educational system.

### 1.1.2. Theories of innovations

One of the most widely used theories used to explore the phenomenon of innovations, the Diffusion of Innovation, was developed by Everett M. Rogers (Rogers, 1995). An innovation is communicated through certain channels over time among the members of a social system (Rogers, 1995). According to the theory, diffusion is a special process that is related to the spread and scaling of new ideas, and it is unavoidably linked to a certain degree of uncertainty in an individual or organization. The four main elements in the diffusion of new ideas are (1) innovation, (2) communication channels, (3) time, and (4) social system.

The scientists raised the question why certain innovations spread quickly and why some did not. In the opinion of Rogers (2003), every innovation bears certain characteristics that could be grouped into several major attributes such as (1) relative advantage, (2) compatibility, (3) complexity, (4) trialability, and (5) observability. The rate of the adoption of an innovation depends on these characteristics.

Relative advantage is the characteristic which defines the degree to which an innovation is perceived as better than the status quo. There is no bigger difference to what extent the advantage is better—most important is whether individuals realize that an innovation responds to their current needs. Compatibility indicates whether an innovation integrates well, and is consistent, with existing systems, standards, norms, values, and experiences. Complexity is a characteristic of innovation that shows whether innovation is perceived as easy to apply in daily practice. Trialability indicates how much of experimentation does an innovation allow for. Observability defines visibility, which, in other words, is the result of the innovation implemented.

Innovations that individuals perceive as having greater relative advantage, compatibility, trialability, observability, and less complexity will be adopted more rapidly than other innovations.

Diffusion is essentially a social process through which people talking to people spread an innovation (Rogers, 1995). The decision process is a mental process that consists of five stages:

1. Knowledge
2. Persuasion
3. Decision
4. Implementation
5. Confirmation.

In the Knowledge stage, a person becomes aware of an innovation and realizes how it works or functions. The Persuasion stage happens when a position or an attitude toward an innovation is developed; it may be either favorable or unfavorable. The next stage is Decision; this is the stage where a person decides to accept or to reject an innovation. During the Implementation stage, the accepted innovation is put into practice, and the implementation of innovation thus commences at this point. The final stage is Confirmation; this is the stage where the results of an innovation-decision are evaluated.

Rogers (1995) also contributed to the development of understanding of different types of innovators, and accordingly, he grouped the innovators under the following categories: (1) innovators, (2) early adopters, (3) early majority, (4) late majority, and (5) laggards. This categorization is based on the percentage of individuals (or organizations) under each portion of the normal curve, marked off by standard deviations from the mean (Rogers, 2002).

1. Innovators make up 2.5% of the individuals in a system to adopt an innovation. They are defined as individuals having more cosmopolite social relationships.
2. Early adopters make up 13.5% of the population that adopts an innovation. Early adopters are considered locals, but they have strong leadership in most systems in relation to the adoption and implementation of innovation.
3. Early Majority and Late Majority adopters comprise the apex of the adoption bell curve with each category comprising 34% of the population. Early Majority adopters usually adopt innovation if they are convinced that the whole process is relatively easy and does not create high pressure.
4. The last adopters, Laggards, make up for 16% of the population, and they are more traditional and more suspicious about innovations; thus, it takes much longer for this group of individuals to adopt innovations.

Rogers' theory of the diffusion of innovations (1995) is widely used in investigating the spread of innovations in different contexts, including in education. Very often, it is used while analyzing technological innovations in education (Sahin, 2006). The theory is also used in investigating teachers' innovative practices and innovative behavior in secondary education and other levels of the education system. For example, Curtis (2020) aimed to discover if high school geography teachers' decisions to use geospatial technologies conformed to Rogers' (2003) theory and the progress of stages of adoption of an innovation. The study revealed that knowledge construction occurred simultaneously through actions in the Persuasion, Implementation, and Confirmation stages. This study presented a different process of decision-making that does not adhere to a linear timeline.

Grgurović (2014) analyzed an application of the diffusion of innovations theory for the investigation of blended language learning. The results received from the qualitative data (in-depth teacher interviews, class and lab observations, and a student focus group) and quantitative data (student surveys) indicated that the innovation (blended language learning) shared common attributes with other innovations and that both positive and negative innovation attributes were present. Timucin (2009) also addressed the theory of diffusion of innovations for exploring the diffusion of technological innovations in teaching and learning foreign languages.

There are other theories that help explore the phenomenon of innovations. One of such theories is called the theory of social change. Most innovations, according to this theory (Coser, 1957), happen because of the social conflict, which changes traditions, norms, and systems such those in economic and technological realms.

The theory of Social Practices analyzes how newly developed social practices change existing practices by preserving, replacing, or disrupting interrelated connections between practices (Reckwitz, 2003).

Koleck (2014) addresses the theory of Social Networks while analyzing innovations. Social networks influence processes of learning, socialization, and innovation. Implementation and diffusion of innovations are social processes, and, thus, social networks play a major role in the spread and adoption of innovations. Through the larger network of social relations, social networks may impact whether and how innovations are accepted. This idea is also supported by Schröder and Krüger (2019, p. 25), who note that "the connection to local networks can hence explain how social innovation in education can successfully be implemented."

Towndrow et al. (2010) draw attention to other approaches to investigate and explore innovations. They refer to the framework of the complexity of innovations proposed by Cohen and Ball (2007). Researchers (Towndrow et al., 2010) note that the following features must be considered when evaluating whether an innovation was successfully implemented at scale: adopters, innovations, and environments. Cohen and Ball (2007) define adopters as individuals or organizations, for example, schools, which adopt innovations and the relationships between innovators and adopters as well as the resources of adopters. Innovations is a multilayered feature and includes sponsorship (e.g., grants), targeted area/sector of education (e.g., assessment system), relationship to existing practices, standards, level of ambition and elaboration, and scaffolding of the innovation. Environments are institutional/organizational contexts that influence the users of innovation.

## 1.2. The concept of innovative work behavior

*The section focuses on the concept of innovative work behavior as a process. This section describes the peculiarities of innovative work behavior at different stages of innovation according to Rogers' Diffusion theory.*

The scaling of innovations is also related to the concept of innovative work behavior. Researchers agree that teachers' innovative work behavior plays a very crucial role in enhancing the performance of schools and society, and it should be the main concern of teachers (Thurlings et al., 2015).

In order to better understand the meaning of the concept of innovative work behavior, one needs to remember the core work roles of an employee. Welbourne (1998) suggested four core work roles for an employee: the roles of the worker, organizational citizen, team player, and career achiever. The implementation of innovations require extra role behavior that is necessary for organizations to survive (Tuominen & Toivonen, 2011). Innovations in organization require innovative work behavior. "Innovative work behavior can be understood as a series of behaviors in which individuals generate novel ideas or schemes in their work, solve practical problems at work and achieve positive effects, including the generation, development and implementation of innovative ideas and so on" (Sun & Huang, 2019, p. 571).

Farr and Ford (1990) define innovative work behavior as an individual's behavior that aims to achieve the initiation and intentional introduction of new ideas, products, or procedures. Jansen (2004) defines innovative work behavior as "the intentional creation, introduction, and application of new ideas within a work role, group or organisation, in order to benefit role performance, the group or the organisation" (Janssen, 2000, p. 288). This kind of innovative behavior has three dimensions, such as idea generation, idea promotion, and idea realization (Janssen, 2004). Kleysen and Street (2001) and Hosseini and Shirazi (2021) define innovative work behavior as actions aimed at the generation, introduction, and application of innovations (novelty) within an organization.

Innovative work behavior is reflected by a series of behaviors in which individuals generate novel ideas, solve practical problems at work, and achieve positive effects (Sun & Huang, 2019, p. 571). Janssen (2000) highlighted three important types of innovative behavior: (a) generating new ideas, (b) disseminating one's own ideas and those of others throughout the organization, and (c) working to implement one's own ideas and those of others. The three types of innovative behavior are distinct from one another (Axtell et al., 2000; Zhou & George, 2001). For instance, although a teacher with extensive professional experience is expected to come up with new ideas in education regularly, they may be less expected to be involved in implementing those ideas, which is often a matter of concern for junior teachers. The dimensions of innovative behavior seem clear enough, which include the generation, development, and implementation of innovative ideas. However, there is no consensus in the scientific literature on the concept and dimensions of innovative work behavior. Various scholars (Janssen, 2003; Noefer et al., 2009; Borasi & Finnigan, 2010; De Jong & Hartog, 2010; So, 2013) have highlighted the different dimensions of innovative work behavior.

Realizing educational innovation requires understanding the meaning of innovation. According to Brewer and Tierney (2012), innovation has two components: the new idea and the change that results from the adoption of the new idea. "Creativity is thinking up new things. Innovation is doing new things" (Levitt, 2002). Amabile et al. (1996) state that "all innovation begins with creative ideas." The ability to generate new ideas is not enough for the successful implementation of innovation in practice. People who are full of new ideas often do not understand how to implement those ideas in practice (Levitt, 2002). Innovative work behavior is rather a broad concept involving creativity, which is very important in the generation of new ideas as well as in the application of those ideas (Gilson & Litchfield, 2017).

According to Scott and Bruce (1994), innovative behavior not only includes generating ideas but also the adoption and implementation of those ideas or solutions (Scott & Bruce, 1994). De Jong and Hartog (2010) state that innovative work behavior differs from employee creativity and is more related to the production of new and useful ideas concerning products, services, and processes. Creativity can be seen as a crucial component of innovative work behavior. De Jong and Hartog (2010) distinguish four dimensions of innovative work behavior: idea exploration, idea generation, idea championing, and idea implementation.

Thurlings and colleagues (2015) analyze the concept of innovative work behavior in the educational field and describe it as self-initiated actions, such as bringing forth, developing, applying, promoting, or modifying new ideas (Thurlings, Evers, & Vermeulen, 2015).

The concept of innovative work behavior is based on psychological models of creativity and innovation (Amabile, 1996; West & Farr, 1990), and it encompasses all work activities involved in developing innovative products and processes (Scott & Bruce, 1994). Messman and Mulder (2014) noticed that the existing conceptualization of innovative work behavior did not adequately take into account the fact that innovation processes are characterized by the dynamic relations among the prerequisite innovation tasks and are bound to their particular work context. Relatively few studies focus specifically on teachers' innovative behavior and its determinants (Klaeijsen et al., 2018; Thurlings et al., 2015). Studies about teachers' innovative work behavior point out the role of self-efficacy (Runhaar, 2008), role of work engagement (Konermann, 2011), creative requirements (Binnewies & Gromer, 2012), and openness, motivation, job satisfaction, and interaction within the job (Messman & Mulder, 2011). Different factors may contribute to innovative work behavior of science teachers.

Our study aims to reveal the role of personal and organizational factors in the innovative work behavior of science teachers. We aim to increase the understanding of the impact of personal and organizational factors on science teachers' innovative work behavior while taking into account the interrelatedness of those factors.

## 1.3. Psychological background of innovative work behavior

*This section analyzes innovative work behavior on the basis of three psychological theories: Social Learning theory, Social Cognitive theory, and Expectancy-value theory. In this section, we discuss the personal, cognitive, and environmental factors of innovative work behavior.*

"Nowadays teachers are not only responsible to teach students about knowledge, skills and attitudes according to standard syllabus determined by the Ministry, but also encouraged to be innovative in their teaching" (Aziah & Al Amin, 2018, p. 246).

In pursuit of its research objectives, this study focuses on the discussion and justification regarding the innovative work behavior of science teachers. For this reason, the present research illuminates an individual's innovative work behavior by using a psychological approach. Considering this, the three theories highlighted at the start of this section—Social Learning theory (SLT), Social Cognitive theory (SCT), and Expectancy-value theory (EVT)—are applied to understand the innovative work behavior of science teachers.

SLT explains the behavior of individuals and states that individuals do not automatically observe the behavior of a model and imitate it (Fig. 1.3.1). There are some mediational processes: attention (for a behavior to be imitated, it has to capture our attention), retention (how well the behavior is remembered), reproduction (ability to perform the behavior that the model has just demonstrated), and motivation (rewards and punishment; Bandura, 1977b). We can explain the innovative work behavior of science teachers based on SLT and Roger's diffusion theory. For Rogers (2003), the innovation decision process involves five stages: (1) knowledge, (2) persuasion, (3) decision, (4) implementation, and (5) confirmation. At the first stage of the innovation decision process, an individual is looking for the innovation, and the questions "What?"; "How?"; and "Why?" become critical at this stage. At the knowledge stage, the individual attempts to determine "what the innovation is and how and why it works" (Rogers, 2003, p. 21). This stage corresponds to the attention process according to SLT (Fig. 1.3.2). In this stage, the innovation must grab individual's attention.

SLT states that the retention process follows the attention process. Innovative behavior occurs when the example of new behavior is remembered. The retention process encompasses the persuasion and decision stages, according to Rogers' diffusion theory (2003). At the persuasion stage, the individual remembers the examples of new behavior and, forming an opinion, develops a negative or positive attitude toward innovation. "The formation of a favorable or unfavorable attitude toward an innovation does not always lead directly or indirectly to an adoption or rejection" (Rogers, 2003, p. 176). At the decision stage, the individual chooses to adopt or reject the innovation. While adoption refers to the "full use of an innovation as the best course of action available," rejection means the decision of "not to adopt an innovation" (Rogers, 2003, p. 177).

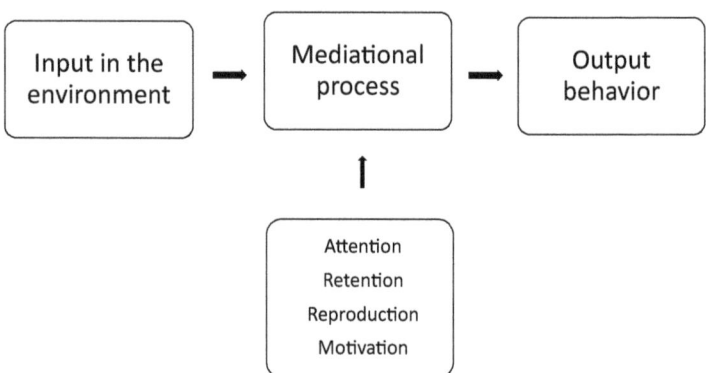

**Fig. 1.3.1:** The modeling of individual's behavior: Case of Social Learning theory

According to SLT, the reproduction process starts after the retention process. During the reproduction process, the individual demonstrates the ability to perform the behavior that the model has just demonstrated. The reproduction process corresponds to the implementation process of Rogers' diffusion theory. At the implementation stage, an innovation is put into education. Rogers (2003) states that uncertainty about the outcomes of the innovation still can be a problem at this stage. At the confirmation stage, the individual looks for support for their decision. According to Sahin (2006), the individual tends to stay away from these messages and seeks supportive messages that justify their decision. The messages work like rewards and punishment and can motivate or demotivate an individual. It means that the last stage of Rogers' diffusion theory of innovations, the confirmation, corresponds to the motivation process of SLT (Fig. 1.3.2). The messages arrive from outside, from the external environment.

SLT explains how innovative work behavior only includes internal processes: attention, retention, reproduction, and motivation. However, SLT cannot explain a very important component of Rogers' diffusion of innovations theory: communication channels. A channel is how an individual relates to their environment and gets messages from the others. Communication occurs through channels. Communication is "a process in which participants create and share information with one another in order to reach a mutual understanding" (Rogers, 2003, p. 5).

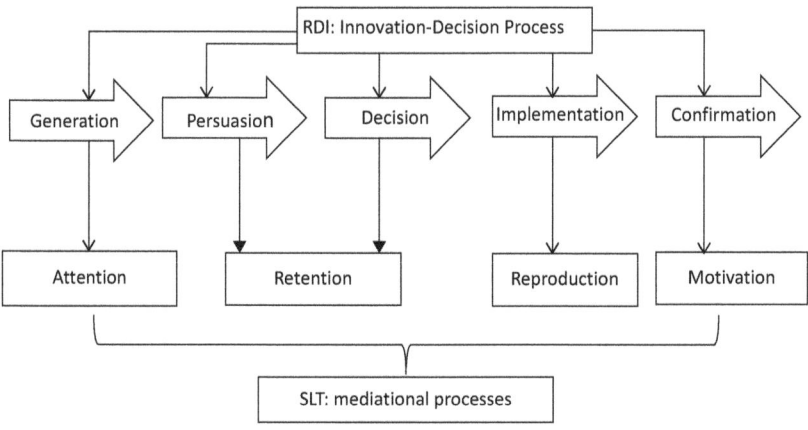

**Fig. 1.3.2:** Innovative work behavior: Case of Social Learning theory and Diffusion theory

Social Cognitive theory (SCT) takes into the consideration the social influence over behavior, including both internal and external social reinforcements in the process of adopting and maintaining an innovation (Fig. 1.3.3; Bandura, 1986; Naslund et al., 2019). According to SCT, there are three main factors that function in a reciprocal triadic determinism: person (cognition), behavior, and environment (Clark & Zimmerman, 2014; Wallin et al., 2018). SCT defines human behavior as an interaction between personal factors, behavior, and the environment (Pajares, 2002). Janssen (2004) highlights the fit between SCT and innovative behavior, emphasizing the *intentional* creation, introduction, and application of new idea.

SCT addresses the cognitive beliefs underlying agentic behavior, which also involves an intentional change in the external environment (Bandura, 1977, 1997, 2001).

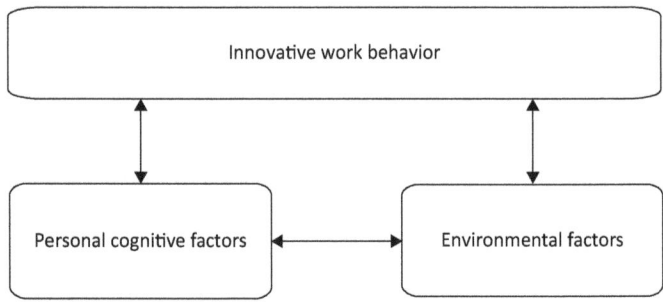

**Fig. 1.3.3:** Innovative work behavior: Case of Social Cognitive theory

Environmental factors also motivate an individual to go for the implementation of innovations. Innovations can only be successfully developed if employees engage in social interactions and reflect on their innovative contributions and on the process of innovation development (Messmann & Mulder, 2012, 2014). Messman and Mulder state that "teachers' engagement [is low] in innovation tasks with low target specificity, while perceived social support for innovation enhanced vocational teachers' engagement in innovation tasks with high target specificity" (Messman & Mulder, 2014, p. 95). "When challenged with such an innovation, the teaching staff first and foremost perceives it and reacts in an affective and cognitive way—resulting, for example, in openness and acceptance or refusal and denial regarding the innovation" (Teerling et al., 2020, p. 11).

SCT suggests that individuals hold beliefs about their ability to make things happen through their own actions (also known as self-efficacy); those self-efficacy beliefs determine behavioral intensity (Bandura, 2012). Self-efficacy is a key element of SCT. Self-efficacy refers to a belief in one's own capabilities to attain a goal (Bandura, 1997a).

Expectancy-value theory (EVT) is then a bridge between SLT and SCT. On one hand, EVT encompasses one component of SCT, that is, self-efficacy, and, on the other, one component of SLT, that is, motivation (Fig. 1.3.4). We can explain science teachers' motivation for innovation on the basis of EVT.

EVT states that the motivation of individual's behavior comes from the combination of their needs and the value of their goals (Eccles et al., 1983; Wigfield, & Eccles, 2000). The goal is a component of SCT and depends on the personal cognitive factor (Fig. 1.3.3). According to EVT, science teachers' motivation for innovations is determined by two factors: expectancies for success of innovations, and subjective task values (Fig. 1.3.4). Expectancies for success refer to how confident a science teacher is about their ability to succeed in an innovative task in the future, either in the short term or in the long term. Whereas task values refer to how important the science teacher perceives the innovative task to be. Expectancies and task values interact to predict main outcomes in science teachers' motivation for innovation (Nagengast et al., 2011; Trautwein et al., 2012).

According to EVT, expectation for success in the innovation activities of teachers depends on their self-confidence about working on innovations (Fig. 1.2.4). This construct is an important point of reference in psychological theories of motivation (Bandura, 1994; Eccles, 2009; Sheldrake, 2016). Bandura (1994) stated, "People with high assurance in their capabilities approach difficult tasks as challenges to be mastered rather than as threats to be avoided" (Bandura, 1994, p. 2). In contrast, people who doubt their capabilities "dwell on their personal deficiencies, on the obstacles they will encounter, and all kinds of adverse outcomes rather than concentrate on how to perform successfully" (Bandura, 1994, p. 2).

The self-confidence phenomenon can be expanded according to various fields of activity. Self-confidence appears to be beneficial within academic area (science, social science, etc.; Sheldrake, 2016). Self-confidence encompasses academic self-concept (Jansen et al., 2014) and self-efficacy (Bong & Skaalvik, 2003; Jansen et al., 2015; Fig. 1.3.4). The SCT suggests that self-efficacy beliefs determine behavioral intensity, particularly when the domains of those beliefs and the type of behavior in question are in alignment with one another (Bandura, 2012). Furthermore, SCT takes account of an individual's past experience, as it acts as an influential component that could influence their motivations, reinforcements, and expectations to become a part of, and give rise to, early precaution behavior. Klaeijsen and colleges (2018) reveled that both intrinsic motivation and occupational self-efficacy serve as predictors for innovative behavior. The relationship between intrinsic motivation and innovative behavior is significant but weak, whereas the relationship between occupational self-efficacy and innovative behavior is much stronger (Klaeijsen et al., 2018). Konermann (2011) found that teachers' occupational self-efficacy mediates the positive relationship between teachers' work engagement and their innovative work behavior.

Ng and Lucianetti (2016) analyzed the role of organizational factors to explain why and how individuals became motivated to make innovations happen on the basis of SCT. They propose that organizational trust and perceived respect by colleagues promote creativity, persuasion, and change in the self-efficacy of employees. Ng and Lucianetti (2016) state that when employees have increasing trust in their organizations, they feel increasingly confident about promoting innovation. Ng and Lucianetti (2016) revealed the role of organizational trust in the self-efficacy of employees: Organizational trust promotes creative self-efficacy and increases generation of new ideas; it increases persistence in efforts to improve self-efficacy and increases the dissemination of new ideas; it increases the use of self-efficacy in bringing in desired changes and promotes the implementation of new ideas.

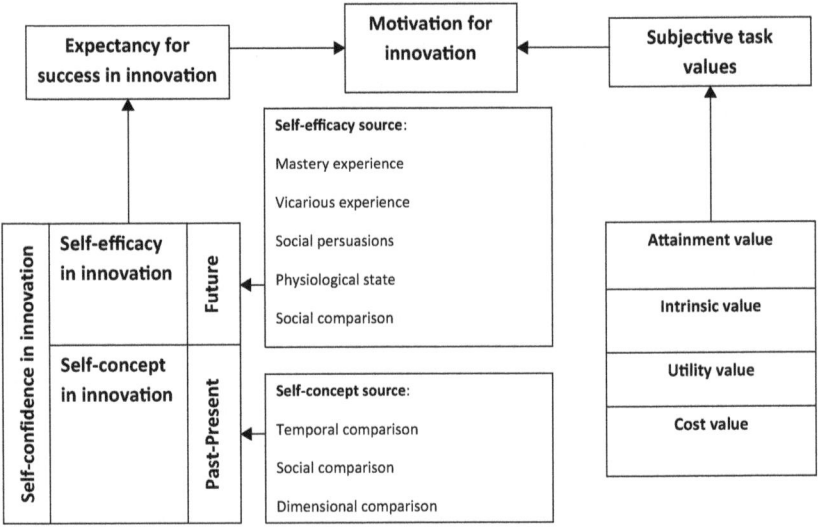

Fig. 1.3.4: Motivation for innovations: Case of Expectancy-value theory

*In this study, we relied on Social Cognitive theory and examined the role of organizational commitment and organizational leadership on the innovative behavior*

*of science teachers. In addition, we analyzed the links between personal cognitive factors and innovative behavior of science teachers. We relied on Social Learning theory, analyzing the role and influence of sociodemographic factors on science teachers' innovative work behavior. Expectancy-value theory was also important to us, helping us to reveal the essence of science teachers' self-confidence and to discover the associations between science teachers' self-confidence and innovative behavior.*

# 2. Methodology

## 2.1. Basic paradigms of science teachers' innovative work behavior

The research presented in this monograph is based on philosophical commitments (positivism and constructivism) and psychological theories (Social Learning theory, Social Cognitive theory, Expectancy-value theory). "Learning theories should be adjusted in a time in which knowledge is no longer acquired in linear manner, technology performs many of the cognitive operations previously performed by learners" (Mattar, 2010, p. 10). We described the relationship between learning theories and innovative work behavior in the theoretical part of the monograph. In this section, we provide a brief description of philosophical commitment (positivism and constructivism).

Positivism confirms the role of science and mathematics in the process of cognition. The main postulates of positivism paradigm are as follows: (1) Objective reality exists that can be known; (2) reality is inherently ordered (Comte, 1957; Durkheim, 1982). All knowledge regarding matters of fact is based on data of experience, and beyond the realm of fact exist pure logic and mathematics. Druckman and Donohue (2020) state that positivism is rooted in assumptions of linear causation, and the goal of empirical research is to achieve internal validity of research results. We accepted the assumptions of a positivist worldview using different statistical techniques suitable for analyzing data on innovative work behavior. We used a categorically independent variable (IV) on a scaled outcome, that is, dependent variable (DV), such as innovative work behavior of science teachers. We accepted the assumptions of positivist worldview ensuring internal validity of research data. We also attempt to address issues of external validity of research embedding randomized sample surveys. Positivism emerges in our research through the acceptance of assumptions of linear causation, which is referred to as linear regression and is based on the principle of least squares.

Constructivism centers on how individuals arrive at the knowledge that enables them to cope with the world (Good, Wandersee, & Julien, 1993). Constructivism is a theory of knowledge in the making, a theory of how learners make sense of the world (Good, Wandersee, & Julien, 1993). Constructivist perspective asserts that individuals construct knowledge from within, by engaging in innovative activities such as problem-solving and experiential and enhancing learning.

The fundamental theoretical postulate of constructivism is that the real world is the context for individuals to construct their new knowledge. Beck and Kosnik (2006) state that constructivist teaching, with its emphasis on the creative and innovative process of knowledge acquisition, can be used to improve outcomes and promote positive attitudes and confidence among teachers.

Two forms of constructivism were those proposed by Piaget (1978) and Vygotsky (1978). Jean Piaget emphasizes the importance of the cognitive processes that occur within individuals—"cognitive constructivist" (Osborne & Wittrock, 1983; Piaget, 1978). Cognitive constructivism therefore emphasizes the personal construction of knowledge. Lev Vygotsky emphasizes the importance of society, culture, and language—"social constructivism" (Lemke, 2001; Vygotsky, 1978). Social constructivism has been covered broadly in discussions in the literature concerning teaching and learning. Social constructivism lays emphasis on communication where individuals are given opportunities to interact with others in order to create new meaning (Palmer, 2005). We analyzed the innovative work behavior of science teachers by both forms of constructivism. In line with social constructivism, we analyzed the role of organizational commitment, professional development, and organization leadership support in science teachers' innovative work behavior. In line with cognitive constructivism, we emphasized the role of self-confidence in science teacher's innovative work behavior.

## 2.2. Research methods and the design of research articles

This part of the monograph addresses the methods and instruments constructed and used in the present research. It introduces the underlying principles of those instruments used to gather relevant data for the present research. Our research is based on a quantitative approach, which is a systematic and empirical investigation of the motivation-for-learning phenomenon using statistical or computational techniques. The objective of quantitative research is to develop mathematical models pertaining to innovative work behavior phenomenon. The process of measurement is a basis for quantitative research because it provides the fundamental connection between empirical observations and mathematical relationships.

In the monograph, we perform a secondary analysis of data of the international study Trends in International Mathematics and Science Study (TIMSS 2015). TIMSS 2015 is the sixth assessment in the TIMSS series since its introduction in 1995. TIMSS 2015 not only helps to examine students' achievement but also allows for the analysis of different factors associated with student achievement—home, community, school, classroom environment, and teaching methods. TIMSS 2015 research model consists of various questionnaires designed to address specific areas—Student Questionnaire, Home Questionnaire, Teacher Questionnaire, School Questionnaire, Curriculum Questionnaire. We were only interested in the Teacher Questionnaire to conduct an empirical research into the performance of science teachers targeted in the present study. The science teacher questionnaire asked science teachers about their education, professional development, experience in teaching, and the educational activities and materials used in the class of students selected for the TIMSS assessment. Teacher questionnaires encompass questions about teacher education and training (BTBS 23; BTBS 24), teacher experience (BTBG 01), and new education technology and methods used in the science classroom. Hooper (2016) states that "Effective teachers are able to create an optimal classroom environment by providing clear purpose and 'strong guidance' for the classroom while encouraging cooperation among the students and an environment of respect between students as well as between students and the teacher" (Hooper, 2016, p. 78).

An effective teacher is involved in innovative activities. TIMSS 2015 provides a unique opportunity to analyze the innovative work activities of science teachers by posing the following questions:

*BTBG 09E—How often do you have the following types of interactions with other teachers? Work together to try out new ideas.*

*BTBG 09D—How often do you have the following types of interactions with other teachers? Visit another classroom to learn more about teaching.*

*BTBG 14F—How often do you do the following in teaching this class? Ask students to decide their own problem-solving procedures?*

*BTBG 14G—How often do you do the following in teaching this class? Encourage students to express their ideas in class.*

*BTBG 09C—How often do you have the following types of interactions with other teachers? Share what I have learned about my teaching experiences.*

It should be noted that these questions were removed from the latest TIMSS 2019 survey questionnaire, and we did not have the opportunity to analyze the innovative behavior of science teachers on the basis of the latest TIMSS 2019 database.

In this case, for secondary data analysis, we used only those questions of TIMSS 2015 that addressed science teachers' innovative work behavior. The secondary data analysis of answers received for these questions is performed in each article of the third chapter of our monograph—Empirical Insight of Factors and Actors of Science Teachers Innovative Work Behavior. The articles of the third chapter (see the table of content of the present monograph) of monograph examine the innovative behavior of science teachers and the factors driving their innovative behavior. The structure of all the articles in the empirical chapter of the monograph is the same: introduction, methodology, results, discussion, and conclusions.

The relevance of the quantitative study is substantiated in the Introduction section of articles. Introduction is derived from a review of theoretical literature sources. In the Introduction part of the article, the main parameters of quantitative research are presented: research problem, research object, and research purpose. All literature sources mentioned in the Introduction section are included in the general bibliography, which is provided at the end of the monograph in the Bibliography section.

The methodology is unique for each article. It is determined by the purpose of the article. The methodology part describes the research questions, their scales, samples, and justifies the choice of mathematical statistics. In addition, an initial analysis of the data in terms of its suitability for mathematical statistics is performed in the methodology part of the article. We checked the normality of data, exceptions of data, and removed those exceptions.

In the Results section of the article, we present a discussion on the statistical analysis of TIMSS 2015 data carried out for the present study. The choice of statistical analysis is determined by the purpose of the study. In the Discussion section, we discuss the empirical research carried out for the present study and present its main findings. Moving forward, we provide our conclusions drawn from the findings of the present study.

In accordance with the research ethics set out in TIMSS 2015, we present our acknowledgment of the data source. SOURCE: TIMSS 2015 Assessment Frameworks. Copyright © 2013 International Association for the Evaluation of Educational Achievement (IEA). Publisher: TIMSS & PIRLS International Study Center, Lynch School of Education, Boston College.

# 3. Empirical insight into factors and actors driving science teachers' innovative work behavior

## 3.1. The internal structure of science teachers' innovative work behavior

*This section analyzes the internal structure of science teachers' innovative work behavior using the confirmatory factor analysis and path analysis. Using The Trends in International Mathematics and Science Study (TIMSS 2015) data on Lithuanian science teachers as the basis, we created a hierarchical linear modeling based on the tenets of diffusion theory: (1) knowledge, (2) persuasion, (3) decision, (4) implementation, and (5) confirmation.*

### 3.1.1. Introduction

Innovations are essential for a society to progress. Innovation is defined as "a new or improved product or process (or combination thereof) that differs significantly from the unit's previous products or processes and that has been made available to potential users (product) or brought into use by the unit (process)" (Vincent-Lancrin et al., 2019, p. 17). "Innovations in education are of particular importance because education plays a crucial role in creating a sustainable future" (Serdyukov, 2017, p. 5). Education not only needs new ideas and inventions but also their successful and effective implementation to enhance the quality of both teaching and learning (Shelton, 2011).

Scholars have distinguished two aspects of innovation: "product" innovation and "process" innovation (Vincent-Lancrin, 2019; Halász, 2018). Product innovation means teachers create "different products and services, such as new syllabi, textbooks or educational resources, or new pedagogies or educational experiences" (e.g., e-learning or new qualifications). Process innovations on the other hand are demonstrated by teachers' innovative work behavior; for example, "how teachers work together, how they group students and manage other aspects of their learning experience; they may collaborate with other entities, use new marketing and external relations methods, new forms of communication with students and parents, etc. In the case of services such as education, products and processes may also be difficult to tell apart" (Vincent-Lancrin et al., 2019).

Vincent-Lancrin et al. (2019) state that while it is easy to talk about innovation in education, it is difficult to talk about implementation of innovations in educational practice (p. 3). There are many sources that provide guidance on the different ways of measuring innovations in educational practice, including the following: the New Consortium Media (Adams Becker et al., 2018); *A New Perspective* (OECD, 2014); An Innovation Survey (Halász, 2018); Olso Manual (2018). Various international research databases are available that provide useful guidance in studying innovation: "We use the Programme on International Student Assessment (PISA), Trends in International Mathematics and Science Study (TIMSS) and Progress in International Reading Literacy Study (PIRLS) databases to cover and identify these key practices at the classroom or school levels" (Vincent-Lancrin et al., 2019). Using Rogers' (2003) Diffusion theory as the premise of our research, we utilized the TIMSS (2015) data on Lithuania to explore science teachers' innovative work behavior.

Previously published research has proven that innovative behavior is defined as a self-initiated, three-stage process: (a) intentional idea generation, (b) idea promotion, and (c) idea realization (Janssen, 2003; Thurlings et al., 2015). Thurlings et al. (2015) noticed that some authors did not explore all of Janssen's (2003) stages. Noefer et al. (2009) did not distinguish the second stage of idea promotion. Borasi and Finnigan (2010) focused on the first and third stages. So (2013) only focused on the first stage (idea generation). This study aims to bridge the above-mentioned three-stage process of science teachers' innovative behavior.

The aim of this study is to reveal the system and structure of science teachers' innovative work behavior on the basis of the TIMSS 2015 data.

### 3.1.2. Method of research

A secondary analysis of the TIMSS 2015 data analysis performed according to the theoretical model of science teachers' innovative behavior. The TIMSS 2015 instrument for science teachers allowed us to carry out an empirical analysis of science teachers' innovative behavior. We used three questions from the TIMSS 2015 questionnaire about science teachers' innovative behavior: BTBG 09C (How often do you have the following types of interactions with other teachers? Share what I have learned about my teaching experiences); BTBG 09D (How often do you have the following types of interactions with other teachers? Visit another classroom to learn more about teaching); BTBG 09E (How often do you have the following types of interactions with other teachers? Work together to try out new ideas). According to diffusion theory, the answers to these question revealed teachers' innovative behavior in the science teaching activity (sharing new ideas, development of new ideas in education, visiting another classroom in order to learn more about teaching, generating of new ideas working together). We analyzed science teachers' activity in science classroom by the aspect of applying and promoting of new ideas on the basis of two questions: BTBG 14F (How often do you do the following in teaching this class? Ask students to decide their own problem-solving procedures); BTBG 14G (How often do you do the following in teaching this class? Encourage students to express their ideas in class; Tab. 3.1.1).

Tab. 3.1.1: TIMSS 2015 questions about science teachers' innovative behavior

| Innovative behavior abilities | Question code | Question content |
| --- | --- | --- |
| Generation of new ideas in education | BTBG 09E | How often do you have the following types of interactions with other teachers? Work together to try out new ideas |
| Development of new ideas in education | BTBG 09D | How often do you have the following types of interactions with other teachers? Visit another classroom to learn more about teaching |
| Applying of new ideas in education | BTBG 14F | How often do you do the following in teaching this class? Ask students to decide their own problem-solving procedures |
| Promotion of new ideas in education | BTBG 14G | How often do you do the following in teaching this class? Encourage students to express their ideas in class |
| Modification and sharing of *new ideas* | BTBG 09C | How often do you have the following types of interactions with other teachers? Share what I have learned about my teaching experiences |

The normality of these questions was checked using the Kolmogorov–Smirnov test and values of asymmetry (skewness and kurtosis; Tab. 3.1.2). The results of the Kolmogorov–Smirnov test and the values of asymmetry indicate that the data do not meet the normality conditions (Tab. 3.1.2). The values for asymmetry (skewness and kurtosis) between −2 and +2 are considered to be adequate to prove normal univariate distribution (George & Mallery, 2010). The kurtosis values many times exceed the allowed asymmetry limit. It was investigated which questionnaire data do not meet asymmetry conditions. The Box plot methodology was used to determine which values of the study fall into the yellow and red areas. The questionnaires (959, 956, 957, 223, 593, 646, 474, 929, 643, 473) that entered into the yellow and the red zone were removed. It influenced the sample of the investigation. The survey sample comprised 937 respondents.

Tab. 3.1.2: Results of Kolmogorov–Smirnov test and the values of asymmetry

|  | Generation of new ideas | Development of new ideas | Applying of new ideas | Promotion of new ideas | Modification and sharing of new ideas |
|---|---|---|---|---|---|
|  | BTBG 09E | BTBG 09D | BTBG 14F | BTBG 14G | BTBG 09C |
| Skewness | 2.453 | 1.787 | 1.795 | 2.218 | 1.464 |
| Std. Error of Skewness | .079 | .079 | .079 | .079 | .079 |
| Kurtosis | 19.626 | 17.423 | 17.284 | 15.300 | 1.136 |
| Std. Error of Kurtosis | .158 | .158 | .158 | .158 | .158 |
| Kolmogorov–Smirnov test | .000 | .000 | .000 | .000 | .000 |

The normality of data was checked after removing exceptions. Asymmetry coefficients indicate that the data satisfy the condition of normality (Tab. 3.1.3).

Tab. 3.1.3: Values of data asymmetry

|  | Generation of new ideas | Development of new ideas | Applying new ideas | Promotion of new ideas | Modification and sharing of new ideas |
|---|---|---|---|---|---|
|  | BTBG 09E | BTBG 09D | BTBG 14F | BTBG 14G | BTBG 09C |
| Valid | 947 | 947 | 947 | 947 | 947 |
| Skewness | -.261 | -.621 | -.628 | -.476 | 1.463 |
| Std. Error of Skewness | .079 | .079 | .079 | .079 | .079 |
| Kurtosis | -.516 | .915 | .443 | -.515 | 1.131 |
| Std. Error of Kurtosis | .159 | .159 | .159 | .159 | .159 |

## 3.1.3. Results of research

Studies of innovative behavior abilities of science teachers were carried out by paying attention to aspects such as generation of new ideas in education, development of new ideas in education, confirmation and sharing of *new ideas* (Tab. 3.1.4). The respondents evaluated their abilities on a four-point scale: "Very often"; "Often"; "Sometimes"; or "Never or almost never." The results revealed that science teachers gave a higher score to sharing of *new ideas* that they have learned about teaching experience than other innovative behavior activities: often—47.0%; very often—10.6% (Tab. 3.1.4).

Tab. 3.1.4: Innovative work behavior (generation-development-sharing) of science teachers: percentage frequency

| Innovative behavior | Question code | Question content: How often do you have the following types of interactions with other teachers? | Very often | Often | Sometimes | Never or almost never |
|---|---|---|---|---|---|---|
| Generation of new ideas | BTBG 09E | Work together to try out new ideas | 3.7 | 20.9 | 65.8 | 9.6 |
| Development of new ideas | BTBG 09D | Visit another classroom to learn more about teaching | 5.2 | 27.3 | 61.8 | 5.7 |
| Confirmation and sharing of new ideas | BTBG 09C | Share what I have learned about my teaching experiences | 10.6 | 47.0 | 41.1 | 1.4 |

Science teachers less positively evaluated the generation of new ideas in education abilities (Tab. 3.1.4). As much as 20.9% of respondents indicated that they often generate new ideas in education, and only 3.7% respondents indicated that they generate new ideas in education very often. The qualitative comparison of science teachers' innovative behavior abilities revealed that the ability to share *new ideas* was rated twice better (47.0%—often, 10.6%—very often) than the ability to generate new ideas (20.9%—often, 3.7%—very often; Tab. 3.1.4).

The sharing of new ideas represents the essence of Rogers' (2003) Diffusion theory. According Rogers (2003), the diffusion process includes these communication elements: an innovation, two individuals or other units of adoption, and a communication channel. The diffusion is "the process in which an innovation is communicated through certain channels over time among the members of a social system [...] a process in which participants create and share information with one another in order to reach a mutual understanding" (Rogers, 2003, p. 5). The main communication channels are *mass media* and *interpersonal communication*. The TIMSS 2015 instrument about science teachers innovative behavior is based on interpersonal communication: Teachers work in groups generating new ideas; visit another classroom to learn more about teaching. Interpersonal communication is less important in generating new ideas but more important in sharing new ideas (Tab. 3.1.4).

We analyzed science teachers' innovative behavior by the aspect of implementing (applying and promoting) new ideas in the science classroom: BTBG 14F (How often do you do the following in teaching this class? Ask students to decide their own problem-solving procedures); BTBG 14G (How often do you do the following in teaching this class? Encourage students to express their ideas in class; Tab. 3.1.5). A secondary analysis of TIMSS 2015 data showed that 70% of science teachers in every or in almost every lesson encourage students to express their ideas in class (Tab. 3.1.5). This means that science teachers have good abilities in promoting new ideas in the science classroom. A low percentage of science teachers, 13.3%, ask students to decide their own problem-solving procedures in every or in almost every lesson, and 32.6% of science teachers ask students to decide their own problem-solving procedures for about half of the lessons. These data prove the teachers' strong ability to apply innovations in the practice of science lessons. A qualitative comparison of science teachers' ability to generate and share new ideas (Tab. 3.1.4) and the ability to implement new ideas in science education (Tab. 3.1.5) revealed that science teachers' abilities to implement new ideas are stronger than their abilities to generate new ideas.

**Tab. 3.1.5:** Innovative work behavior of science teachers in classroom: percentage frequency

| Innovative behavior | Question code | Question content: How often do you do the following in teaching this class? | Every or almost every lesson | About half the lessons | Some lessons | Never |
|---|---|---|---|---|---|---|
| Applying of new ideas in education | BTBG 14F | Ask students to decide their own problem-solving procedures | 13.3 | 32.6 | 50.7 | 3.4 |
| Promotion of new ideas in education | BTBG 14G | Encourage students to express their ideas in class | 70.0 | 23.1 | 6.8 | 0.1 |

The data on science teachers' innovative work abilities were analyzed on the basis of diffusion theory. According to this theory, the innovation-decision process involves five steps: (1) knowledge, (2) persuasion, (3) decision, (4) implementation, and (5) confirmation (Rogers, 2003). The TIMSS 2015 instrument for science teachers allowed for analyzing science teachers' innovative work behavior activities at the stages of idea generation (knowledge, persuasion, and decision), implementation, and confirmation (Tab. 3.1.6). A comparative analysis of the highest ranks percentages sum confirmed that science teachers' idea-generating abilities (first—second—third stage, according to Diffusion theory) are weaker than their abilities in the implementation and confirmation of new ideas (Tab. 3.1.6).

Tab. 3.1.6: Science teachers' innovative work behavior abilities: sum of percentages of two highest ranks (very often, often/every or almost every lesson, about half the lessons)

| Science teachers' innovative behavior abilities | Question code | Stages of innovations according to diffusion theory | | |
|---|---|---|---|---|
| | | The idea generation stage: knowledge, persuasion, decision | The implementation Stage | The confirmation Stage |
| Generation of new ideas | BTBG 09E | 23.7 | | |
| Development of new ideas | BTBG 09D | 32.5 | | |
| Applying of new ideas in education/ problem-solving | BTBG 14F | | 45.9 | |
| Promotion of new ideas in education/ encourage students to express new ideas | BTBG 14G | | 93.01 | |
| Confirmation and sharing of *new ideas* | BTBG 09C | | | 57.6 |

The difference between science teachers' innovative work behavior abilities was statistically assessed. We used the Freedman criterion because the study samples were dependent and the question scales were ranked. A statistically significant difference was detected between all science teachers' innovative behavior abilities. A nonparametric Freedman test of difference among repeated measures was conducted and rendered a $\chi^2$ value of 1803,419, which was significant ($p < 0.01$).

We analyzed the correlation between science teachers' innovative abilities (Tab. 3.1.7). We used Spearman correlation in order to assess whether the questions' scales were ordinal and data were nonparametric.

Tab. 3.1.7: Spearman correlation between science teachers' innovative work behavior components

|  | Sharing of new ideas | Development of new ideas | Try out new ideas | Applying new ideas | Promotion of new ideas |
|---|---|---|---|---|---|
| Sharing of new ideas | 1,000 | ,463** | ,452** | ,186** | ,155** |
| Development of new ideas |  | 1,000 | ,588** | ,224** | ,106** |
| Generating new ideas |  |  | 1,000 | ,240** | ,119** |
| Applying new ideas |  |  |  | 1,000 | ,253** |
| Promotion of new ideas |  |  |  |  | 1,000 |

**. Correlation is significant at the 0.01 level (two-tailed).

A statistically significant correlation was determined between science teachers' ability to try out new ideas and looking for new ideas (i.e., visiting other classroom; $r = 0.588^{**}$, $p = 0.01$); the ability to look for new ideas and share new ideas ($r = 0.463^{**}$, $p = 0.01$); the ability to try out new ideas and share new ideas ($r = 0.452^{**}$, $p = 0.01$; Tab. 3.1.7). On one hand, this means teachers can come up with new ideas for teaching by attending other teachers' classes. On the other, statistically significant correlation coefficients confirm that teachers can discover new educational ideas by sharing their new teaching experiences with other teachers.

Two correlations between applying of new ideas in education are weak but statistically significant: looking for new ideas ($r = 0.224^{**}$, $p = 0.01$); generating new ideas ($r = 0.240^{**}$, $p = 0.01$; Tab. 3.1.7). This result can be explained by the fact that new ideas can be applied in education only when they are discovered. New ideas in education are being discovered by practical (looking for new ideas) and mental (generating of new ideas) sources. The correlation coefficient between the generation and application of new ideas in education ($r = 0.240^{**}$, $p = 0.01$) is slightly higher than the correlation coefficient between looking for new ideas and application of new ideas ($r = 0.224^{**}$, $p = 0.01$). Fisher r-to-z transformation for dependent samples did not confirm this difference.

A statistically insignificant correlation was determined between science teachers' ability to promote new ideas and all innovative abilities (except the ability applying new ideas in education; Tab. 3.1.7).

## Results of path analysis: Full model

Using the path analysis procedure with AMOS 17, we tested the theoretical model of science teachers' innovative behavior abilities (Fig. 3.1.1). The theoretical model was carried out on the basis of diffusion theory (Rogers, 2003). Path analysis confirmed that the structural model fits the data well (Tab. 3.1.8).

In reference to model fit, researchers use goodness-of-fit indicators to check the model. According to Benter (1990), the measurement model fits the data well if:

- $\chi^2/df$ is less than 5, describes the distance between model and data, but depends on sample size;
- RMSEA (Root Mean Square Error of Approximation) is less than 0.08, describes how much error or unexplained variance remains after fitting model;
- NFI (Normed Fit Index) is larger than 0.80;
- IFI (Incremental Fit Index) exceeds 0.90;
- TLI (Tucker–Lewis Index; also Non-normed Fit Index [NNFI]) exceeds 0.90, describes the "power" of model compared to "the situation without the model";
- CFI (Comparative Fit Index) exceeds 0.90, describes the "power" of model compared to "the situation without the model" (Bentler, 1990).

**Tab. 3.1.8:** Fitness of items of science teachers' innovative behavior abilities

|  | Absolute fit index | | | Relative fit index | | |
|---|---|---|---|---|---|---|
|  | $\chi^2/df$ | RMSEA | GFI | IFI | TLI | CFI |
| Assumed model | 0.701 | .000 | .877 | 1.000 | 1.002 | 1.000 |
| Acceptance value | 1–5 | <.08 | >.80 | >.90 | >.90 | >.90 |

The explanatory power of science teachers' innovative behavior model (Fig. 3.1.1) was assessed by calculating the coefficient of determination ($R^2$) of endogenous constructs (BTBG 09D, BTBG 14F, BTBG 14G, BTBG 09C; Tab. 3.4.4.8). The data indicate that 51.5% of the variation in science teachers' ability to develop new ideas scores can be accounted for by their ability to generate new ideas in education. Also, 63.3% of the variance of science teachers' ability to share new teaching ideas can be accounted for by their ability to promote new ideas in science education and encouraging students to express their ideas in classroom. The smallest $R^2$ value was detected by checking whether the development of new ideas affects science teachers' ability to apply new ideas in problem-solving processes in classroom ($R^2 = 0.086$). It means only 8.6% of the variance in science teachers' ability to apply new ideas in problem-solving processes in classroom can be accounted for by their ability to develop new ideas about science teaching.

The main purpose of this study was to reveal the system and structure of science teachers' innovative behavior abilities in intentional idea generation, idea promotion, and idea realization. We examined both the direct and indirect effects for significance and magnitudes (Tab. 3.4.4.8). We found that the indirect effect of science teachers' ability to develop new ideas on their ability to share new teaching ideas (BTBG 09D →BTBG 09C) did not correspond to the model and hence this path was deleted. All direct paths were significant in the final model (Fig. 3.1.1). The most significant effects in the model include the following: the effect of promotion of new ideas in science education and encouraging students to express their ideas in classroom (BTBG 14G) on the sharing of new teaching ideas (BTBG 09C; $B = 4.252$); the effect of applying new ideas in problem-solving processes (BTBG 14F) on science teachers' ability to promote new ideas in education (BTBG 14G; $B = 0.352$); the effect of generating new ideas about teaching (BTBG 09E) on science teachers' ability to develop new ideas (BTBG 09D; $B = 0.603$; Fig. 3.1.1).

# Structure of innovative work behavior

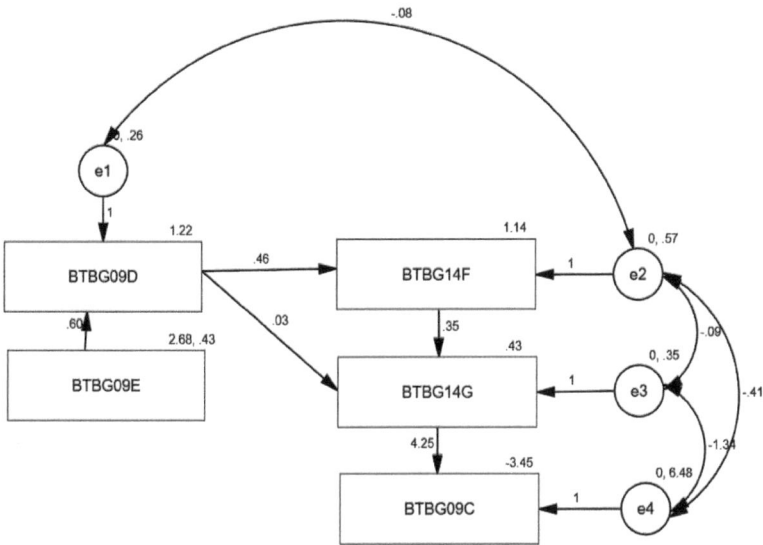

**Fig. 3.1.1:** The structure of science teachers' innovative work behavior: unstandardized coefficients

**Tab. 3.1.9:** The structure and system of science teachers' innovative behavior abilities: direct and indirect path coefficients and statistical significance

| Hypothesis | Paths | Path coefficients | $p$ | $R^2$ | Results | Effect |
|---|---|---|---|---|---|---|
| H1. Generating new ideas directly affects science teachers' ability to develop new ideas | BTBG 09E → BTBG 09D | .703 | <.001 | .515 | Support | Direct |
| H2. Development of new ideas directly affects science teachers' ability to apply new ideas in the problem-solving processes in the classroom. | BTBG 09D → BTBG 14F | .557 | <.001 | .086 | Support | Direct |

*(continued on next page)*

Tab. 3.1.9: Continued

| Hypothesis | Paths | Path coefficients | p | $R^2$ | Results | Effect |
|---|---|---|---|---|---|---|
| H3. Applying new ideas in the problem-solving process affects science teachers' abilities to promote new ideas in education | BTBG 14F → BTBG 14G | .782 | <.001 | .316 | Support | Direct |
| H4. Promotion of new ideas in education affects science teachers' ability to share new ideas. | BTBG 14F → BTBG 09C | .992 | <.001 | .633 | Support | Direct |
| H5. Development of new ideas indirectly affects science teachers' ability to promote new ideas in education. | BTBG 09D → BTBG 14G | 0.022 | 0.739 | 0.321 | Against | Indirect |

### 3.1.4. Discussion

This study aims to explore science teachers' innovative work behavior on the basis of Diffusion theory (Rogers, 2003). For the purposes of the present study, we used the TIMSS 2015 data on science teachers from Lithuania and analyzed that data using a hierarchical linear modeling that was based on the components of diffusion theory—that is, (1) knowledge, (2) persuasion, (3) decision, (4) implementation, and (5) confirmation (Rogers, 2003). We did not use any specialized or customized questionnaires. We performed a qualitative analysis of TIMSS 2015 questions in order to create a group of questions to explore the innovative work behavior of science teachers. As mentioned early, the TIMSS 2015 instrument for science teachers allowed for the analysis of science teachers' innovative behavior abilities at the stages of idea generation (knowledge, persuasion, decision), implementation, and confirmation.

The purpose of this study is to investigate the system and structure of science teachers' innovative work behavior in intentional idea generation, idea promotion, and idea realization on the basis of TIMSS 2015 data. Our analysis revealed that science teachers' idea generating abilities (the first, second, and third stages, according to diffusion theory) are weaker than the new ideas implementation in educational practice and new ideas sharing abilities (Tab. 3.1.6).

Hsiao et al. (2011) states that "teachers usually have good ideas, but they still need to discuss with their peers or supervisors, and they even attempt to convince each other. Once in a while, when the teachers' new idea is adopted, the realization of an innovative idea has begun. As a result, it may help teachers create higher innovative work behavior step by step" (Hsiao et al., 2011, p. 34). We found that science teachers' ability to implement new ideas is stronger than their ability to generate new ideas. It can be assumed that science teachers have the ability to discuss with their peers or supervisors implementing new ideas and convince each other.

Marth et al. (2018) analyzed the relationship between science motivation, technology interest, and the innovative work behavior of science teachers in summer science school. Marth et al. (2018) discuss inquiry-based learning methods like innovative learning and believe that science teachers' innovative behavior can be a powerful tool for promoting school students' motivation in science studies and enhancing their interest in technology. Furthermore, they observed that if "teachers are highly motivated and interested in science as well as in technology by furthermore using new methods like IBSE approaches in the classroom, a spill over to students seems likely" (Marth et al., 2018, p. 58). We analyzed how science teachers apply innovation in the classroom by asking students to decide their own problem-solving procedures. The results of our study revealed that 45.9% of science teachers very often or often ask students to decide their own problem-solving procedures (Tab. 3.1.5). It was surprising that a vast majority of science teachers, 93.01%, encourage in every or in almost every lesson or in about half the lessons encouraged students to express their new ideas in classroom. But we didn't analyze how science teachers' innovative behavior influences students' motivation for learning science, interest in technology, and the innovative behavior of students. Other scholars found that "knowledge application of teachers' knowledge innovation has positive moderation effect on students' creativity self-efficacy and students' innovation behaviors. Knowledge generation has negative moderation effect on students' creativity self-efficacy and students' innovation behaviors" (Chang, 2018, p. 1).

Using Rogers' diffusion theory (Rogers, 2003) as the basis of our study, we carried out a statistical analysis of science teachers' innovative behavior abilities by using a path analysis procedure with AMOS 17. Rogers (2003) justified the innovation-decision process by communication channels. The scheme of direct communication channel is as follows: Knowledge→Persuasion→Decision→Implementation→Confirmation (Rogers, 2003). Our empirical model of science teachers' innovative behavior abilities that was developed based on a path analysis procedure with AMOS is as follows: Generating of new ideas in teaching science→Development of new ideas→Applying of new ideas in education→Promotion of new ideas in education→Sharing new ideas about teaching science (Tab. 3.1.6). We also analyzed one indirect way: Development of new ideas indirectly affects science teachers' ability to promote new ideas in education (Fig. 3.1.1). The paths in our empirical model (Fig. 3.1.1) correspond to the channels in Rogers' (2003) model. Our path analysis with AMOS confirmed the Diffusion theory. All direct paths were significant in the final model (Fig. 3.1.1). The ability of science teachers to generate new ideas in teaching science is statistically significant in affecting their ability to develop new ideas, apply and promote new ideas in education, and share new ideas about teaching science. As we mentioned previously, the most significant effects in the model include the following: the effect of promotion of new ideas in science education encouraging students to express their ideas in classroom on sharing new teaching ideas ($\beta = 0.992$); the effect of applying of new ideas in problem-solving processes on science teachers' abilities to promote new ideas in education ($\beta = 0.782$); and the effect of generating new ideas about teaching on science teachers' ability to develop new ideas ($\beta = 0.703$).

### 3.1.5. Conclusions

Science teachers of Lithuania apply innovations in practice. The comparative analysis of the highest ranks' percentages sum confirmed that science teachers' idea-generating ability (from the first to the third stage, according to diffusion theory) is weaker than their ability to implement and confirm new ideas. A statistically significant difference was detected between all science teachers' innovative behavior abilities.

A statistically significant correlation was determined between science teachers' ability to try out new ideas and looking for new ideas while discussing with others, look for new ideas and share new ideas, and try out new ideas and share new ideas.

The path analysis based on the TIMSS 2015 data of science teachers' innovative work behavior components confirmed the Diffusion theory. The effect of promotion of new ideas in science education by encouraging students to express their ideas in classroom upon sharing new teaching ideas was mostly significant. The effect of applying of new ideas in problem-solving processes on science teachers' abilities to promote new ideas in education was big and statistically significant. The effect of generating of new ideas about teaching on science teachers' ability to develop new ideas was also statistically significant.

## 3.2. Comparative analysis of demographic and educational factors on science teachers' trying out and sharing new ideas

> *In this section, we analyze two innovative activities: trying out new ideas and sharing new ideas. We performed ordinal logistic regression analysis of TIMSS 2015 data on Lithuanian teachers and tested the role of three factors (gender, level of education, and age of science teacher) in science teachers' activities in relation to trying out and sharing new ideas.*

### 3.2.1. Introduction

Successful changes in the education system are due to the innovative behavior of teachers. What are the factors driving teachers' innovative behavior? In scientific literature, the answer to this question is often sought in the professional development of teachers (Runhaar, 2008; Shen, Benson & Huang, 2014; Van der Heijden et al., 2015; Kaleem et al., 2018). Runhaar (2008) analyzed what factors determine teachers' professional development through of Human Resources Management (HRM) in schools and how teachers' professional development can be explained and promoted. Scholars (Runhaar, 2008; Kaleem et al., 2018) came to the conclusion that HRM can serve as a tool in professional development. It should be noted that Runhard (2008) tried to explain professional development in terms of innovative behavior and knowledge sharing within teams.

On one hand, one has to notice that teachers' professional development can encourage their innovative behavior. On the other, both professional development and innovative behavior are determined by teachers' demographic factors. This most noticeable group of identifiers—demographic factors—should not be overlooked when studying the driving factors behind science teachers' innovative behavior. A review of the literature has revealed the main demographic factors that influence the innovative behavior of science teachers': age, gender, years of teaching experience, years of education, and the level of education (Carmeli et al., 2006; Runhaar, 2008; Yang and Huang, 2008; Thurlings et al., 2015).

Scholars (Thurlings et al., 2015) have analyzed a wide range of quantitative studies examining the influence of demographic factors on teachers' innovative behavior and concluded that "The demographic variables that were explored in the quantitative studies were gender (one study), age (two), income (one), teaching experience (two), years of education (one), level of education (one), tenure (one), and other functions (e.g., ICT coordinator, literacy coach) or other tasks within the teaching profession (one). Only the variables, income, years of education, and having other functions had a significant positive effect on innovative behavior" (Thurlings et al., 2015, p. 14).

An analysis of scholarly literature has revealed a lack of scientific insights into science teachers' innovative behavior and especially about the role of demographic factors on science teachers' innovative work behavior. Using the TIMSS 2015 data on science teachers of Lithuania, an ordinal logistic regression (OLR) analysis was carried out to explore the relationship between science teachers' abilities to work together to try out new ideas in education and their demographic factors (age, gender, years of teaching experience, years of education, and level of education). The aim of this study is to lead to a better understanding of the innovative behavior of science teachers through correlating their innovative behavior with demographic factors.

### 3.2.2. Method of research

We used OLR for data analysis, which is a special case of generalized linear modeling. Ordinal regression is used with ordinal dependent variables. Whereas independent (predictor) variables can be categorical factors or continuous covariates. Ordinal regression models can be called "cumulative logit models" (like a variant of logistic regression). Ordinal regression models are also called "proportional odds models" because they generate proportional (parallel) regression lines. It means ordinal regression models share the same regression coefficients which vary at the point where they intercept. It means for each level of dependent variable we have the same $b$ coefficients but different intercept "threshold" in SPSS.

The mathematical model of OLR analysis is described by several equations. Suppose dependent variable (Y) depends on three interval or ordinal regressors (predictors) X, Z, W. Three mathematical models for the logarithms (logit functions) of the dependent variable are created:

$$ln\frac{P(Y \leq 1)}{P(Y > 1)} = C_1 - b_1 X - b_2 Z - b_3 W \qquad (1)$$

$$ln\frac{P(Y \leq 2)}{P(Y > 2)} = C_2 - b_1 X - b_2 Z - b_3 W \qquad (2)$$

$$ln\frac{P(Y \leq 3)}{P(Y > 3)} = C_3 - b_1 X - b_2 Z - b_3 W \qquad (3)$$

We are aware that in all equations the regression coefficients $b_1$, $b_2$, $b_3$ at the regressors are the same, but the constants $C_1$, $C_2$, $C_3$ (intercepts) are different. A positive coefficient ($b_1$, $b_2$, $b_3$) for a variable indicates that as this variable increases, the likelihood of Y gaining greater values increases. The negative coefficient ($b_1$, $b_2$, $b_3$) of a variable indicates that as this variable increases, the probability of Y gaining greater values decreases. The coefficient ($b_1$, $b_2$, $b_3$) values represent how much the logarithm of the odds ratio will change as the value of the corresponding independent variable increases by one unit when the remaining independent variables are fixed.

In a good data model:

- $\chi^2$ criterion of maximum likelihood is $p < 0.05$.
- Wald criterion for all regressors is $p < 0.05$.
- $\chi^2$ is $p \geq 0.05$ for straight parallelism test.

On the basis of the mathematical model, equations (1), (2), (3), we tested the influence of demographic factors on science teachers' innovative work behavior.

We tested using OLR the TIMSS 2015 data on sciences teachers in Lithuania. The OLR analysis was performed according to Rogers' (2003) Diffusion theory. As previously mentioned (see Section 3.1), the TIMSS 2015 instrument for science teachers allowed for carrying out an empirical analysis of their innovative work behavior. We checked all questions for their fitness in OLR. We selected only two questions from the TIMSS 2015 questionnaire about science teachers' innovative work behavior: BTBG 09C (How often do you have the following types of interactions with other teachers? Share what I have learned about my teaching experiences); BTBG 09E (How often do you have the following types of interactions with other teachers? Work together to try out new ideas). According to Diffusion theory, answers to these questions revealed teachers' innovative work behavior in science teaching activities (sharing new idea, developing new ideas in education by visiting another classroom to learn more about teaching, and generating new ideas by working together). In this study, we describe the results of our OLR analysis for these two questions only.

### 3.2.3. Results of the research

We tested four models of science teachers for two innovative work abilities: trying out new ideas in interactions with other teachers and sharing new ideas on the basis of TIMSS 2015 data of Lithuania.

(I) First model:

TONI = $f$(EDU, GEN, AGE)

Whereas TONI is the ability to try out new ideas, GEN = teacher's gender, AGE = teacher's age, and EDU = level of formal education completed.

(II) Second model:

TONI = $f$(EDU, GEN, AGE, TE)

Whereas TONI is the ability of generation new ideas, GEN = teacher's gender, AGE = teacher's age, EDU = level of formal education completed, and TE = teaching experience according to the duration of the professional activity.

(III) Third model:

SHNI = $f$(EDU, GEN, AGE)

Whereas SHNI is the ability of sharing new ideas, GEN = teacher's gender, AGE = teacher's age, and EDU = level of formal education completed.

(IV) Fourth model:

SHNI = $f$(EDU, GEN, AGE, TE)

Where SHNI is the ability of sharing new ideas, GEN = teacher's gender, AGE = teacher's age, EDU = level of formal education completed, and TE = teaching experience.

The alternative models test four hypotheses (Tab. 3.2.1).

**Tab. 3.2.1:** The hypothesis testing about science teachers' innovative work behavior: TONI and SHNI

| Hypothesis | |
|---|---|
| H1 | The ability of science teachers in trying out new ideas in interactions with other teachers is better predicted by the first model |
| H2 | The ability of science teachers in trying out new ideas in interactions with other teachers is better predicted by the second model |
| H3 | The ability of science teachers in sharing new ideas is better predicted by the third model |
| H4 | The ability of science teachers in sharing new ideas is better predicted by the fourth model |

In the first model, that is, TONI = $f$(EDU,GEN, AGE), we observe the variables EDU (level of formal education completed), GEN (gender), and AGE (age) have an influence over the ordinal variable TONI (science teachers' ability to try out new ideas and generate new ideas in interactions with other science teachers); thus, TONI is dependent on EDU, GEN, and AGE. In our study the ordinal variable TONI had five categories. For the ordinal dependent variable TONI, five categories and four (k-1) equations were created, each with a different intercept but with the same $b$ coefficients (slopes) for predictor variables. It means the effects of independent variables are the same for each level of the dependent variable. We tested this condition with the "test of parallel lines assumption," which is important for the OLR analysis.

In our first and third models, the predictor ("locational") variables are as follows:

EDU (level of formal education completed): 1 = "Did not complete Upper secondary"; 2 = "Upper secondary"; 3 = "Post-secondary, non-tertiary"; 4 = "Short-cycle tertiary"; 5 = "Bachelor's or equivalent"; 6 = "Master's or equivalent"; 7 = "Doctor or equivalent."
GEN (teacher's gender): 1 = female, 2 = male.
AGE (teacher's age): 1 = "Under 25"; 2 = "25–29"; 3 = "30–39"; 4 = "40–49"; 5 = "50–59"; 6 = "60 or more."

In our second and fourth models, the predictor (locational) variables are as follows:

EDU (level of formal education completed): 1 = "Did not complete Upper secondary"; 2 = "Upper secondary"; 3 = "Post-secondary, non-tertiary"; 4 = "Short-cycle tertiary"; 5 = "Bachelor's or equivalent"; 6 = "Master's or equivalent"; 7 = "Doctor or equivalent."
GEN (teacher's gender): 1 = female, 2 = male.
AGE (teacher's age): 1 = "Under 25"; 2 = "25–29"; 3 = "30–39"; 4 = "40–49"; 5 = "50–59"; 6 = "60 or more."
TE (continuous variable) = number of years the teacher has been teaching.
This study measures five constructs with models TONI = $f$(EDU, GEN, AGE) and SHNI = $f$(EDU, GEN, AGE) and six constructs with models TONI = $f$(EDU, GEN, AGE, TE) and SHNI = $f$(EDU, GEN, AGE, TE).

Before we start looking at the effects of each explanatory variable in different models, it is important to determine whether the model improves the possibility to predict the outcome. We compare a model that doesn't include any explanatory variables (the baseline or "Intercept Only" model) with a model that includes all the explanatory variables (the "Final" model—this would normally have several explanatory variables, but, at the moment, it contains only the level of education). We compare the final model against the baseline to see whether it has significantly improved the fit to the data (Tab. 3.2.2). The parameters of the model for which the model fit are calculated (Tab. 3.2.2).

The significant $\chi^2$ statistic ($p < 0.05$) indicates that the final model demonstrates a significant improvement over the baseline, intercept-only, model. The small $p$ value from the model fitting test, $p < 0.05$, indicates that at least one of the regression coefficients in the model is not equal to zero (Tab. 3.2.2).

**Tab. 3.2.2:** Model fitting information derived from the ordinal regression analysis: to explain science teachers' ability to try out new ideas and share new ideas in science education

| Dependent variable | Model | Model | −2 Log Likelihood | $\chi^2$ | df | Sig. |
|---|---|---|---|---|---|---|
| TONI-1 | GNI = f (EDU, GEN, AGE) | Intercept only | 151,798 | | | |
| | | Final | 96,097 | 55,701 | 6 | .000 |
| TONI-2 | GNI = f (EDU, GEN, AGE, TE) | Intercept only | 1111,086 | | | |
| | | Final | 1057,107 | 53,979 | 7 | .000 |
| SHNI-1 | SHNI = f (EDU, GEN, AGE) | Intercept only | 280,408 | | | |
| | | Final | 265,346 | 15,062 | 6 | .020 |
| SHNI-2 | SHNI = f (EDU, GEN, AGE, TE) | Intercept only | 1152,034 | | | |
| | | Final | 1135,578 | 16,456 | 7 | .021 |

The next parameter in the output is the goodness-of-fit table (Tab. 3.2.3). This table contains Pearson's $\chi^2$ statistic for the model. These statistics are intended to test whether the observed data are consistent with the fitted model. If we do not reject this hypothesis ($p > 0.05$), then we can conclude that the data and the model predictions are similar and that we have a good model.

**Tab. 3.2.3:** Goodness-of-fit test: to explain science teachers' ability to try out new ideas and share new ideas in science education

| Dependent variable | Model | $\chi^2$ | df | Sig. |
|---|---|---|---|---|
| TONI-1 | TONI = f (EDU, GEN, AGE) | Pearson 124,650 | 122 | ,416 |
|  |  | Deviance 91,385 | 122 | ,983 |
| TONI-2 | TONI = $f$ (EDU, GEN, AGE, TE) | Pearson 318,404 | 268 | ,019 |
|  |  | Deviance 346,707 | 268 | ,001 |
| SHNI-1 | SHNI = $f$ (EDU, GEN, AGE) | Pearson 112,999 | 126 | ,790 |
|  |  | Deviance 114,599 | 126 | ,758 |
| SHNI-2 | SHNI = $f$ (EDU, GEN, AGE, TE) | Pearson 1415,599 | 1101 | ,000 |
|  |  | Deviance 807,875 | 1101 | ,000 |

Goodness-of-fit test revealed that the models TONI-2 and model SHNI-2 do not fit very well ($p < 0.05$). We eliminated these models and continued to explore only two models with three predictor variables (EDU, GEN, AGE): TONI-1 and SHNI-1.

We calculated the pseudo $R^2$ for two models: TONI-1 and SHNI-1. These are three pseudo $R^2$ values: Cox and Snell, Nagelkerke, and McFadden. Nagelkerke pseudo $R^2$ value is mostly used in ordinal regression. Here (Tab. 3.2.4), the pseudo $R^2$ values indicate that education level, gender, and age explain a relatively small proportion of variation between science teachers' ability to generate new ideas and ability to share new ideas. This means different characteristics can impact science teachers' ability to generate and share new ideas.

**Tab. 3.2.4:** Pseudo $R^2$ Nagelkerke: to explain science teachers' ability to try out new ideas and share new ideas in science education

| Dependent variable | Model | Pseudo $R^2$ Nagelkerke |
|---|---|---|
| TONI-1 | TONI = f (EDU, GEN, AGE) | 0,21 |
| SHNI-1 | SHNI = $f$ (EDU, GEN, AGE) | 0,19 |

**Tab. 3.2.5:** Parameter estimates to explain science teachers' ability to try out new ideas in science education (model TONI-1)

|  |  | Estimate | Std. Error | Wald | df | Sig. | 95% Confidence Interval | |
|---|---|---|---|---|---|---|---|---|
|  |  |  |  |  |  |  | Lower Bound | Upper Bound |
| Threshold | [TONI = 1] | -1,784 | ,779 | 5,240 | 1 | ,022 | -3,312 | -,257 |
|  | [TONI = 2] | ,512 | ,770 | ,443 | 1 | ,006 | -,997 | 2,021 |
|  | [TONI = 3] | 4,092 | ,787 | 27,029 | 1 | ,000 | 2,550 | 5,635 |
|  | [TONI = 4] | 6,831 | ,924 | 54,671 | 1 | ,000 | 5,020 | 8,642 |
| Location | GEN | ,830 | ,210 | 15,599 | 1 | ,000 | ,418 | 1,242 |
|  | AGE | -,071 | ,062 | 1,305 | 1 | ,253 | -,194 | ,051 |
|  | [EDU=2] | 1,271 | 1,651 | ,593 | 1 | ,441 | -1,965 | 4,507 |
|  | [EDU=4] | -1,740 | ,811 | 4,605 | 1 | ,032 | -3,330 | -,151 |
|  | [EDU=5] | ,778 | ,685 | 1,288 | 1 | ,256 | -,565 | 2,121 |
|  | [EDU=6] | ,617 | ,687 | ,807 | 1 | ,369 | -,729 | 1,962 |
|  | [EDU=7] | 0[a] | . | . | 0 | . | . | . |

Link function: Logit.
a. This parameter is set to zero because it is redundant.

Estimating the coefficients of ordinal regression model based on Wald statistics (Tab. 3.2.5), we see TONI-1 Wald coefficient has a statistical value of 15.599 and a $p$ value of 0.000 and, hence, $p < 0.05$. We observe a statistically significant confirmation that the probability of science teachers trying out new ideas in science education depends on the gender of science teacher. For the predictor AGE, Wald coefficient has a statistical value of 1.305 and, hence, $p > 0.05$; thus, we get a statistically insignificant confirmation that the probability of generating new ideas depends on the age of science teachers. Wald statistics confirm the role of fourth education level (EDU) of science teachers, only at the fourth level (short-cycle tertiary). The Wald coefficient for EDU-4 is 4.605 and $p$ value 0.032 and, hence, $p < 0.05$. It means the fourth level (short-cycle tertiary) has a statistically significant influence on the probability of science teachers generating new ideas in science education.

The threshold coefficients (TONI-1) represent the intercepts, specifically the point (in terms of a logit) where the science teachers' ability to generate new ideas might be predicted for the two categories (Tab. 3.2.5). The mathematical model for the logarithms (logit functions) of the dependent variable is created on model TONI-1:

## Analysis of demographic and educational factors

$$\ln\frac{P(GNI \le i)}{P(GNI > i)} = \begin{cases} -1,784, \text{ when } i = 1 \\ 0,512, \text{ when, } i = 2 \\ 4,092, \text{ when, } i = 3 \\ 6,831, \text{ when, } i = 4 \end{cases} + 0,830(\text{GEN}) - 1,740(\text{EDU} - 4) \quad (4)$$

In the coefficient (EDU-4), the negative sign indicates that this variable has negative effects on science teachers' ability to generate new ideas in education. We recall that the negative coefficient indicates that as the variable EDU-4's values increase, the more likely it is for the TONI-1 values to decrease. Thus, information that science teachers acquire in short-cycle tertiary education makes it more likely that they will generate new ideas in education.

The results of OLR analysis of data on Lithuanian science teachers revealed that the predictor of the duration of science teachers' teaching experience (TE) has a statistically significant value (Tab. 8).

The predictor variable [EDU = 4 (scale)] in the OLR was found to contribute to the model. The [Ordered log odds (Estimate)] = −1.740, SE = 0.811, Wald $\chi^2(1)$ = 4.605, $p < 0.05$. The estimated odds ratio favored a positive relationship of nearly 1.034-fold [Exp(estimate) = 1.034, 95% CI (0.629–1.698)] for every unit increase of [EDU = 4 (scale)]. It means the possibility of trying out new ideas decreases 1.034 times for every unit increase of [EDU = 4 (scale)].

The "test of parallel lines" compares the ordinal model which has one set of coefficients for all thresholds (null hypothesis) to a model with a separate set of coefficients for each threshold (general). We are led to reject the assumption of proportional odds if the general model gives a significantly better fit to the data than the ordinal (proportional odds) model ($p < 0.05$). The results of parallel lines test confirm the null hypothesis (Tab. 3.2.6) that the model TONI-1 has one set of coefficients.

Tab. 3.2.6: Test of parallel lines: to explain science teachers' ability to try out new ideas and share new ideas in science education

| Model | | -2 Log Likelihood | $\chi^2$ | df | Sig. |
|---|---|---|---|---|---|
| TONI-1 | Null Hypothesis | 237,617 | | | |
| | General | 189,643[a] | 47,974[b] | 30 | ,222 |

The study revealed the role of science teachers' gender in their ability to try out new ideas in education (Tab. 3.2.5). The estimated odds ratio favored a positive relationship of nearly .550-fold [Exp(estimate) = 0.550, 95% CI (0.364–0.831)] for every unit increase of [GEN (scale)]. It means the possibility of trying out new ideas decreases 1.034 times for every unit increase of [GEN (scale)].

The results of the cross-tabulation of dependent variable (TONI-1) and independent variable (GEN) show that female science teachers generate new ideas in education more often than male science teachers do (Tab. 3.2.7). It can be assumed that such a result may be due to the disproportionate ratio of men ($N = 124$) to women ($N = 835$) in the study. However, when examining the distribution of percentage frequencies by different ranks, we found a reverse trend in women's and men's group (Tab. 3.2.7). The highest percentage of women chose the highest rank "Very often." Meanwhile, the highest percentage of men chose the rank "Never or almost never" (Tab. 3.2.7). The difference was confirmed statistically ($\chi^2 = 19{,}593$, $df = 4$, $p < 0.000$).

Tab. 3.2.7: Cross-tabulation of variables: to explain science teachers' ability to try out new ideas in science education (model TONI -1)

| Ranks | | GEN\SEX OF TEACHER | | Total |
|---|---|---|---|---|
| | | Female | Male | |
| Very often | Count | 48 | 1 | 49 |
| | % within | **98,0%** | 2,0% | 100,0% |
| Often | Count | 241 | 21 | 262 |
| | % within | 92,0% | 8,0% | 100,0% |
| Sometimes | Count | 499 | 91 | 590 |
| | % within | 84,6% | 15,4% | 100,0% |
| Never or almost never | Count | 45 | 9 | 54 |
| | % within | 83,3% | **16,7%** | 100,0% |
| Omitted or invalid | Count | 2 | 2 | 4 |
| | % within | 50,0% | 50,0% | 100,0% |
| Total | Count | 835 | 124 | 959 |
| | % within | 87,1% | 12,9% | 100,0% |

The model TONI-1 finally has been proved. In his case, the condition the test of parallel lines was confirmed statistically ($\chi^2 = 47{,}974$, $df = 30$, $p = 0.222$). The reasonably good fit of the chosen model to the data is indicated by the likelihood ratio criterion ($\chi^2 = 55{,}701$, $df = 6$, $p = 0.000$) and the Pearson criterion ($\chi^2 = 124{,}650$, $df = 122$, $p = 0.416$).

Analysis of demographic and educational factors 61

Now we start looking at the effects of each explanatory variable in the SHNI-1 model (Tab. 3.2.8).

From the results of parameter estimate—that for the variables AGE and EDU the Wald coefficient has a $p$ value > 0.05—we get a statistically insignificant result. It means the probability of sharing new ideas in science education doesn't depend on the age and educational level of the science teacher.

Tab. 3.2.8: Parameter estimates: to explain science teachers' ability to share new ideas in science education (model SHNI -1)

|  |  | Estimate | Std. Error | Wald | df | Sig. | 95% Confidence Interval | |
|---|---|---|---|---|---|---|---|---|
|  |  |  |  |  |  |  | Lower Bound | Upper Bound |
| Threshold | [SHNI = 1] | -1,364 | ,756 | 3,251 | 1 | ,031 | -2,846 | ,119 |
|  | [SHNI = 2] | 1,105 | ,754 | 2,147 | 1 | ,043 | -,373 | 2,583 |
|  | [SHNI = 3] | 4,787 | ,791 | 36,612 | 1 | ,000 | 3,236 | 6,338 |
|  | [SHNI = 4] | 6,084 | ,878 | 48,039 | 1 | ,000 | 4,363 | 7,804 |
| Location | GEN | ,509 | ,187 | 7,416 | 1 | ,006 | ,143 | ,876 |
|  | AGE | ,025 | ,058 | ,181 | 1 | ,670 | -,090 | ,139 |
|  | [EDU=2] | ,344 | 1,524 | ,051 | 1 | ,821 | -2,642 | 3,331 |
|  | [EDU=4] | -,806 | ,807 | 1,000 | 1 | ,317 | -2,387 | ,774 |
|  | [EDU=5] | ,230 | ,680 | ,115 | 1 | ,735 | -1,102 | 1,562 |
|  | [EDU=6] | ,031 | ,681 | ,002 | 1 | ,963 | -1,304 | 1,366 |
|  | [EDU=7] | 0[a] | . | . | 0 | . | . | . |

Link function: Logit.
a. This parameter is set to zero because it is redundant.

$$\ln\frac{P(GNI \le i)}{P(GNI > i)} = \begin{cases} -1,784, \text{ when } i = 1 \\ 0,512, \text{ when, } i = 2 \\ 4,092, \text{ when, } i = 3 \\ 6,831, \text{ when, } i = 4 \end{cases} + 0,509(\text{GEN}) \qquad (5)$$

The study revealed the role of science teachers' gender on their ability to share new ideas in education (Tab. 3.2.8). The results of the cross-tabulation of dependent variable TONI-1 and independent variable GEN show that female science teachers generate new ideas in education more often than male science teachers

(Tab. 3.2.9). However, when examining the distribution of percentage frequencies by different ranks, we found a reverse trend in women's and men's group (Tab. 3.2.8). The highest percentage of women chose the highest rank "Very often." Meanwhile, the highest percentage of men chose the rank "Never or almost never" (Tab. 3.2.8). The difference was not confirmed statistically ($\chi^2$ = 6,996, $df$ = 3, $p < 0.702$).

Tab. 3.2.9: Cross-tabulation of variables: to explain science teachers' ability to share new ideas in science education (model SHNI-1)

| Ranks | | | Gen\Sex of Teacher | | Total |
|---|---|---|---|---|---|
| | | | Female | Male | |
| Very often | | Count | 92 | 9 | 101 |
| | | % within | 91.1% | 8.9% | 100.0% |
| | | % of Total | 9.8% | 1.0% | 10.7% |
| Often | | Count | 391 | 50 | 441 |
| | | % within | 88.7% | 11.3% | 100.0% |
| | | % of Total | 41.5% | 5.3% | 46.8% |
| Sometimes | | Count | 331 | 56 | 387 |
| | | % within | 85.5% | 14.5% | 100.0% |
| | | % of Total | 35.1% | 5.9% | 41.1% |
| Never or almost never | | Count | 9 | 4 | 13 |
| | | % within | 69.2% | 30.8% | 100.0% |
| | | % of Total | 1.0% | .4% | 1.4% |
| Total | | Count | 823 | 119 | 942 |
| | | % within | 87.4% | 12.6% | 100.0% |
| | | % of Total | 87.4% | 12.6% | 100.0% |

Summarizing the results of ordinal logistic regression, we can state that only (TONI-1) and (Share-1; Tab. 3.2.10) model confirm our hypothesis.

Tab. 3.2.10: The results of hypothesis testing

| Hypothesis | Confirmation of hypothesis |
|---|---|
| H1 The ability of science teachers to try out new ideas in interactions with other teachers is better predicted by the first model | Confirmed |
| H2 The ability of science teachers to try out new ideas in interactions with other teachers is better predicted by the second model | Not confirmed |
| H3 The ability of science teachers to share new ideas of is better predicted by the third model | Confirmed |
| H4 The ability of science teachers to share new ideas is better predicted by the fourth model | Not confirmed |

## 3.2.4. Discussion

In the present study, we analyzed the influence of science teachers' demographic factors (age, sex) and the education factor on their innovative work behavior abilities (trying out new ideas, sharing new ideas) in science education. We investigated the teachers' innovative work behavior by carrying out a secondary analysis of TIMSS 2015 data. The theoretical background of investigation is framed based on Diffusion theory (Rogers, 2003). For the purpose of the study, we chose quantitative access by applying OLR. It should be noted that researchers are more likely to use quantitative methods when studying the influence of demographic factors on teachers' innovative work behavior. Thurlings et al. (2015) conducted extensive literature analysis on the role of demographic factors in teachers' innovative work behavior and found that "an overview of the demographic factors, of which all except one were found in quantitative research" (Thurlings et al., 2015, p. 14).

Scholars do not very often apply regression analysis to investigate teachers' innovative work behavior (Thurlings et al., 2015). We carried out an OLR analysis to ascertain the influence of science teachers' demographic factors and educational factor on their ability to try out and share new ideas in science education. OLR helped us to predict science teachers' innovative wok behavior abilities with ordinal-level dependent variables (TONI, SHNI) and a set of independent variables (GEN, AGE, EDU). The dependent variable is TONI, and SHNI is the order response category variable in TIMSS 2015.

According to Thurlings et al. (2015), a wide range of demographic factors have been studied: gender, age, income, level of education, years of education, and teaching experience. "Only the variables, income, years of education, and having other functions had a significant positive effect on innovative behavior" (Thurlings et al., 2015, p. 14). There was no income question in the TIMSS questionnaire. For this reason, we were unable to investigate the role of this factor.

The TIMSS questionnaire included a question about years of education. We had included this question in the OLR model. The model was validated in a sample of 959 secondary school science teachers from Lithuania. However, the OLR model revealed that this predictor has no statistically significant influence on science teachers' innovative work behavior abilities (TONI and SHNI). This may have been due to the fact that we were not investigating the innovative work behavior of all teachers but only that of the science teachers. The study should be repeated with TIMSS database for other countries. This would make it clear how the number of years a teacher has been teaching relates to their innovative work behavior.

The most important contribution of our study was to validate a mathematical model that incorporates demographic characteristics as predictors of science teachers' two innovative work behavior abilities (TONI and SHNI) in science education. Regarding factors that had the most influence—the TONI and SHNI innovative work behavior abilities of science teachers—results indicate that level of education is statistically significant and is related with TONI. It must be stated that short-cycle tertiary level (EDU-4; Tab. 3.2.5) negatively influences the science teachers' ability to generate new ideas in science education. An analysis of English teachers' abilities to use new information and communication technology (ICT) features in education carried out by Yang and Huang (2008) showed that higher levels of education and "higher levels of computer training, computer literacy, well-supported school environment [...] result in higher task intensity, impact concerns and more technology-mediated teaching behaviour in the classroom" (Yang & Huang, 2008). It should be noted that researchers (Yang & Huang, 2008) understand the application of ICT as an innovation in educational practice. The results of this research are in line with the work done by Yang and Huang (2008) that the education level variable predicts the TONI ability of science teachers. However, there was no statistically significant influence of education levels on science teachers' ability to share new ideas (SHNI; Tab. 3.2.6).

The study has two main limitations. First limitation is that the TONI-1model is based on the data for a single country; that is, only science teachers from Lithuania were studied. Future studies could replicate this research at the international level by verifying mathematical models on the basis of databases for other countries that participated in TIMSS 2015.

The second limitation of this study is that TONI-1 model doesn't have high pseudo $R^2$ values (e.g., Nagelkerke = 21%). It indicates that EDU, AGE, and GEN explain a relatively small proportion of the variation in science teachers' ability to generate new ideas in education. This was what we had expected from the outset because there are numerous characteristics that can have an impact on science teachers' innovative work behavior. These will be addressed in other sections of this monograph.

### 3.2.5. Conclusions

The following conclusions can be drawn from the mathematical model TONI-1 of this research which was created based on the TIMSS 2015 data: (1) science teachers trying out new ideas collaboratively depends on their level of education and gender; (2) the short-cycle tertiary level (EDU-4) negatively influences science teachers' ability to try out new ideas collaboratively in education; (3) science teachers' gender positively influences their ability to try out and share new ideas in education. The female teachers are more likely to try out and share new ideas than male teachers are.

## 3.3. The role of organizational commitment in the innovative work behavior of science teachers

> *In this section, we analyze the role of environmental factors in the innovative work behavior of science teachers of Lithuania. We use a confirmatory factor analysis (CFA) and structural equation modeling (SEM) analysis in order to determine the influence of organizational/affective commitment (OAC) on science teachers' innovative work behavior.*

### 3.3.1. Introduction

The 21st century is marked by rapid technological advancement and innovations. The innovation abilities of future scientists depend on teachers' innovative behavior in classroom. According to Thurlings et al. (2015), "Innovative behavior can be described as a process in which new ideas are generated, created, developed, applied, promoted, realized, and modified by employees to benefit role performance" (p. 1). Thurging et al. (2015) revealed three main factors influencing the innovative behavior of science teachers: demographic, individual, and organizational. However, Thurlings et al. (2015) state that "little research has been conducted that explores teacher innovative behavior and which factors influence this behavior or what effects can be achieved through such behavior" (p. 1).

Scholars analyzed organizational factors and their effect on innovative work behavior by studying the following features: actors and their relationships with other people such as colleagues and managers (Horng et al., 2005; Borasi & Finnigan, 2010; Mushayikwa & Lubben, 2009); support and feedback (Binnewies & Gromer, 2012; Loogma et al., 2012; Schussler et al., 2007); facilities and resources (Bourgonjon et al., 2013; Donnelly et al., 2011; Nakata, 2011). Khoo, Yeap, and Ramayah (2014) revealed that the process of innovation is enhanced by dynamic environmental conditions.

In summing up the research of other scholars, Thurlings et al. (2015) distinguished the main individual factors and their effect on innovative work behavior: personality, trait, competence, motivation, self-efficacy, persistence, problem-solving, and others. However, there is lack of research about the influence of organizational commitment on the innovative work behavior of teachers, and especially science teachers, whose innovative work behavior in the classroom determines future technological progress in society.

Organizational commitment refers to a teacher's willingness to remain focused and committed in their work (Mowday, Steers & Porter, 1982; Meyer, Stanley, Herscovitch & Topolnytsky, 2002; Abdullah, 2011). According to Meyer and Alen (1997), organizational commitment has three components: affective commitment, continuance commitment, and normative commitment (Meyer et al., 2002). Affective commitment refers to an employees' attachment with their organization; continuance commitment refers to an employee's organizational commitment arising as a result of work relationships and other benefits; normative commitment refers to an employee's sense of obligation that is based on their values and norms.

The aim of this study is to reveal how organizational affective commitment influences the innovative work behavior of science teachers.

### 3.3.2. Method of the research

A secondary analysis of the TIMSS 2015 data was performed using the theoretical model of organizational commitment that comprises the three forms of commitment mentioned previously: affective, continuance, and normative. The TIMSS 2015 instrument for science teachers allowed for carrying out an empirical analysis of science teachers' organizational commitment (BTBG 10 question; Tab. 3.3.1).

Tab. 3.3.1: Science teachers' organizational/affective commitment (based on TIMSS 2015 data)

| Code | Question: How often do you feel the following way about being a teacher? |
|---|---|
| BTBG 10A | I am content with my profession as a teacher |
| BTBG 10B | I am satisfied with being a teacher at this school |
| BTBG 10C | I find my work full of meaning and purpose |
| BTBG 10D | I am enthusiastic about my job |
| BTBG 10E | My work inspires me |
| BTBG 10F | I am proud of the work I do |
| BTBG 10G | I am going to continue teaching for as long as I can |

The skewness and kurtosis were well within a tolerable range for assuming a normal distribution, and an examination of the histograms suggested that the distributions looked approximately normal (Tab. 3.3.2). The values for skewness (asymmetry) and kurtosis between -2 and +2 are considered acceptable in order to prove normal univariate distribution (George & Mallery, 2010).

Tab. 3.3.2: The skewness and kurtosis values of the items of science teachers' organizational/ affective commitment

|  | BTBG 10A | BTBG 10B | BTBG 10C | BTBG 10D | BTBG 10E | BTBG 10F | BTBG 10G |
|---|---|---|---|---|---|---|---|
| Skewness | ,461 | ,639 | ,710 | ,372 | ,241 | ,386 | ,300 |
| Std. Error | ,152 | ,152 | ,152 | ,152 | ,152 | ,152 | ,152 |
| Kurtosis | -,307 | -,072 | -,471 | -,475 | -,802 | -,791 | -,834 |
| Std. Error | ,302 | ,302 | ,302 | ,302 | ,303 | ,303 | ,302 |

### 3.3.3. Results of the research

*Results of Exploratory factor analysis (EFA)*

An EFA was performed on a 7-item TIMSS 2015 scale with varimax (orthogonal) rotation. The Kaiser–Meyer–Olkin test (KMO test) revealed sampling adequacy and KMO = 0.892, $p < 0.05$ for the observed variables (Intercorrelation was checked by using Bartlett's test: $\chi^2(28) = 14{,}981.772$, $p < 0.05$. All elements on the diagonal (Measure of Sampling Adequacy [MSA]) of anti-image correlation matrix are greater than 0.5. Extraction communalities indicate that the variables chosen for this analysis are related with each other. A principal component analysis (PCA) of science teachers' three different forms of organizational commitment explained as much as 65.97% of the variance in the entire set of 7 items. Factor 1 was labeled organizational/emotional commitment due to the high loadings by the following items (Tab. 3.3.3). Tab. 3.3.3 shows the factor loadings for the Affective Commitment factor.

Tab. 3.3.3: Component matrix of organizational/emotional commitment scale. Correlations among the items of the affective commitment scale

| Items code | Items: How often do you feel the following way about being a teacher? | Component 1 |
|---|---|---|
| BTBG 10A | I am content with my profession as a teacher | .879 |
| BTBG 10B | I am satisfied with being a teacher at this school | .877 |
| BTBG 10C | I find my work full of meaning and purpose | .854 |
| BTBG 10D | I am enthusiastic about my job | .809 |
| BTBG 10E | My work inspires me | .805 |
| BTBG 10F | I am proud of the work I do | .768 |
| BTBG 10G | I am going to continue teaching for as long as I can | .664 |

Finally, seven items are combined in one factor—organizational/emotional commitment: I am content with my profession as a teacher, I am satisfied with being a teacher at this school, I find my work full of meaning and purpose; I am enthusiastic about my job; My work inspires me; I am proud of the work I do; I am going to continue teaching for as long as I can (Tab. 3.3.3).

A correlation analysis was performed in order to strengthen the above results of convergence. Spearman correlation was used for this analysis. A correlation analysis was performed for affective commitment items. Table 3.3.4 shows the results of this correlation analysis. Table 3.3.4 clearly indicates that all items correlate in statistically significant ways. The strongest correlation was detected between affective commitment variable BTBG 10D (I am enthusiastic about my job) and variable BTBG 10E (How often do you feel the following way about being a teacher? My work inspires me; $r_s = 0.851$, $p < 0.001$; Tab. 3.3.4). It can be concluded that a statistically significant correlation exists between science teachers' enthusiasm and finding their work inspiring.

**Tab. 3.3.4:** Intercorrelations between the items for affective commitment

|  | BTBG 10A | BTBG 10B | BTBG 10C | BTBG 10D | BTBG 10E | BTBG 10F | BTBG 10G |
|---|---|---|---|---|---|---|---|
| BTBG 10A | 1.000 | .761** | .561** | .626** | .626** | .600** | .506** |
| BTBG 10B |  | 1.000 | .606** | .614** | .613** | .618** | .486** |
| BTBG 10C |  |  | 1.000 | .639** | .598** | .633** | .388** |
| BTBG 10D |  |  |  | 1.000 | .851** | .744** | .530** |
| BTBG 10E |  |  |  |  | 1.000 | .757** | .510** |
| BTBG 10F |  |  |  |  |  | 1.000 | .527** |
| BTBG 10G |  |  |  |  |  |  | 1.000 |

**. Correlation is significant at the 0.01 level (two-tailed).

## Results of Confirmatory factor analysis (CFA)

The organizational/affective commitment (OAC) construct was examined using a CFA. According to Bentler (1990), the measurement model fits the data well if $\chi^2/df$ is less than 5; RMSEA less than 0.08; NFI larger than 0.80; IFI, TLI, and CFI exceed 0.90 (Bentler, 1990). The fitness of items of the OAC factor revealed a sufficient fit and confirmed the 7-item questionnaire's structure (Tab. 3.3.5). The CFA confirmed that the structural model fit the data well). Standardized and unstandardized estimates are shown in Table 3.3.6.

**Tab. 3.3.5:** The fitness of items of the organizational/affective commitment factor

|  | Absolute fit index | | | Relative fit index | | |
|---|---|---|---|---|---|---|
|  | $\chi^2/df$ | RMSEA | GFI | IFI | TLI | CFI |
| Assumed model | 1.640 | .050 | .957 | .980 | .969 | .979 |
| Acceptance value | 1–5 | <.08 | >.80 | >.90 | >.90 | >.90 |

**Tab. 3.3.6:** Standardized and unstandardized coefficients of the organizational/affective commitment factor

| Code of observed variable | Observed variable | Latent construct | B | β | SE | p label |
|---|---|---|---|---|---|---|
| BTBG 10A | I am content with my profession as a teacher | Organizational/ affective commitment (OCE) | .862 | .834 | .137 | <.001 |
| BTBG 10B | I am satisfied with being a teacher at this school | | .859 | .764 | .135 | <.001 |
| BTBG 10C | I find my work full of meaning and purpose | | .859 | .753 | .106 | <.001 |
| BTBG 10D | I am enthusiastic about my job | | 1.026 | .746 | .094 | <.001 |
| BTBG 10E | My work inspires me | | 1.051 | .723 | .095 | <.001 |
| BTBG 10F | I am proud of the work I do | | 1.266 | .607 | .097 | <.001 |
| BTBG1 0G | I am going to continue teaching for as long as I can | | 1.000 | .590 | .137 | <.001 |

The unstandardized beta ($B$) value represents the predictor variables—I am content with my profession as a teacher; I am satisfied with being a teacher at this school; I find my work full of meaning and purpose; I am enthusiastic about my job; My work inspires me; I am proud of the work I do; I am going to continue teaching for as long as I can—and the latent variable—OAC.

With the variable "I am content with my profession as a teacher," for every one-unit increase in the variable "I am content with my profession as a teacher," the dependent variable OAC increases by 0.862 units. The highest $B$ value is of the variable "I am proud of the work I do." For every one-unit increase in the variable "I am content with my profession as a teacher," the dependent variable increases by 1.266 units. It can be stated that the pride that teachers feel about teaching profession mostly influences their organizational commitment.

## Results of Structural modeling of equations (SEM)

We analyzed five hypotheses:

Hypothesis ($H_1$): Science teachers' organizational/affective commitment will have a significant and negative effect on the generation of new ideas in education.

Hypothesis ($H_2$): Science teachers' organizational/affective commitment will have a significant and negative effect on the development of new ideas in education.

Hypothesis ($H_3$): Science teachers' organizational/affective commitment will have a significant and negative effect on the applying of new ideas in education.

Hypothesis ($H_4$): Science teachers' organizational/affective commitment will have a significant and negative effect on the promotion of new ideas in education.

Hypothesis ($H_5$): Science teachers' organizational/affective commitment will have a significant and negative effect on the modification and sharing of new ideas.

Janssen (2003) proposed a three-stage process: (a) intentional idea generation, (b) idea promotion, and (c) idea realization. We accepted this classification of sciences' teachers' innovative work behavior (Tab. 3.3.7). Thus, we state that science teachers' innovative work behavior abilities include the following: generation of new ideas in education, development of new ideas in education, applying new ideas in education, promotion of new ideas in education, and modification and sharing of new ideas (Tab. 3.3.7). Science teachers' innovative work behavior abilities correspond with different stages of the diffusion theory (Tab. 3.3.7).

Tab. 3.3.7: Science teachers' innovative work behavior abilities and the stages of the Diffusion theory

| Innovation behavior abilities | Code of observed variable | Observed variable | The stages of the diffusion theory |
|---|---|---|---|
| Generation of new ideas in education | BTBG 09E | How often do you have the following types of interactions with other teachers? Work together to try out new ideas | The knowledge stage The decision stage The persuasion stage |

*(continued on next page)*

Tab. 3.3.7: Continued

| Innovation behavior abilities | Code of observed variable | Observed variable | The stages of the diffusion theory |
|---|---|---|---|
| Development of new ideas in education | BTBG 09D | How often do you have the following types of interactions with other teachers? Visit another classroom to learn more about teaching | The implementation stage |
| Applying new ideas in education | BTBG 14F | How often do you do the following in teaching this class? Ask students to decide their own problem-solving procedures | |
| Promotion of new ideas in education | BTBG 14G | How often do you do the following in teaching this class? Encourage students to express their ideas in class | |
| Modification and sharing of new ideas | BTBG 09C | How often do you have the following types of interactions with other teachers? Share what I have learned from my teaching experiences | The confirmation stage |

Our hypothesis about sciences teachers' innovative work behavior based on SEM is described graphically (Fig. 3.3.1). We performed an SEM analysis based on data from TIMSS 2015 with AMOS statistical package on the seven questions. The hypothesized model appears to have a good fit with the data. The CFI value represents the overall difference between observed and predicted correlations. A value of 0.979 is situated well over the cutoff value of 0.08. It means the hypothesized model resembles the actual correlations. The TLI is 0.969, and the RMSEA is 0.050 < 0.08 (Tab. 3.3.5). For the good fit of the data to the model, we did not make post hoc modifications.

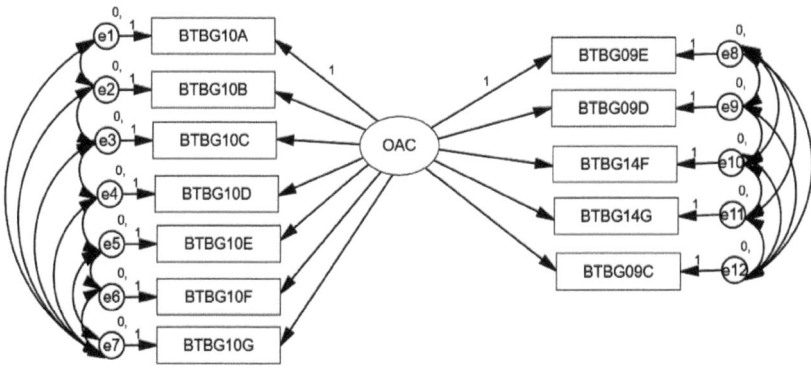

**Fig. 3.3.1:** Structural equation model: science teachers' organizational/affective commitment (OAC) and their innovative work behavior

Results of the direct influence of OAC on science teachers' innovative work behavior abilities were described by unstandardized beta ($B$), standardized beta ($\beta$), squared multiple correlations ($R^2$; Table 3.3.8). The SEM analysis confirmed all hypotheses ($H_1$ to $H_5$; Tab. 3.3.8).

**Tab. 3.3.8:** Results of SEM: influence of organizational/affective commitment on science teachers' innovation behavior

| Code of observed endogenous variable | Innovative work behavior abilities | Observed variable | B | β | SE | $R^2$ | Result |
|---|---|---|---|---|---|---|---|
| | | Direct | | | | | |
| BTBG 09E | Generation | Work together to try out new ideas | .479 | .375 | 0.090 | .141 | Accept |
| BTBG 09D | Development | Visit another classroom to learn more about teaching | .493 | .414 | 0.085 | .171 | Accept |
| BTBG 14F | Applying | Ask students to decide their own problem-solving procedures | .583 | .331 | 0.123 | .110 | Accept |
| BTBG 14G | Promotion | Encourage students to express their ideas in class | .454 | .353 | 0.090 | .125 | Accept |
| BTBG 09C | Sharing | Share what I have learned from my teaching experiences | .509 | .419 | 0.087 | .175 | Accept |

It was found that the OAC had a direct, positive, and statistically significant impact on all the innovative work behavior abilities of science teachers. In particular, it had the biggest influence on science teachers' ability to share innovative ideas ($\beta = 0.414$, $p < 0.05$) and the lowest influence on applying new ideas in education ($\beta = 0.331$, $p < 0.05$).

### 3.3.4. Discussion

The present research examined the influence of OAC on science teachers' innovative work abilities. We adopted the EFA and CFA methods and path analysis to test the structural model. The results indicated that science teachers' OAC had a significant influence on science teachers' innovative work behavior in education. Operationalization of innovative work behavior on the basis of Diffusion theory (Rogers, 2003) revealed group innovative abilities of science teachers: generation of new ideas in education, development of new ideas in education, applying new ideas in education, promotion of new ideas in education, and modification and sharing of new ideas.

Applying new ideas in education is related to problem-solving (Kim et al., 2018). We analyzed the science teachers' ability to ask students to decide their own problem-solving processes. An SEM analysis of the TIMSS 2015 data on science teachers revealed a statistically significant influence of Organization/Affective Commitment on science teachers' ability to ask students to decide their own problem-solving processes. Chang (2018) analyzed the effects of teachers' knowledge innovation on students' innovative work behavior by using a hierarchical linear modeling. It is found that teachers' knowledge innovation has a positive moderation effect on students' innovative work behaviors (Chang, 2018).

Applying new ideas in education is connected to two levels: teacher and student (Chang, 2018). We analyzed teachers' level by Organization/Effective Commitment and analyzed students' level by "asking students to decide their own problem-solving procedures." Chang (2018) analyzed teachers' level by teachers' knowledge innovation and analyzed students' level by students' innovative work behavior. Results of our research about applying new ideas in science education confirm the importance of the two-level model: teacher level and student level.

We analyzed science teachers' innovative work behavior based on relations with other people, such as working together to try out new ideas and visiting another classroom to learn more about teaching. According to Thurlings et al.

(2015), these categories belong to organizational factors (interactions, support, guidance, feedback, and collegiality). We found that OAC had a major influence on science teachers' ability to work together ($\beta = 0.375$, $p < 0.05$) and visiting another classroom to learn more about teaching ($\beta = 0.414$, $p < 0.05$). Our results agree with the findings of other scholars. Messmann (2011) highlights the social aspect of innovative work behavior by stating that the development of innovations is a complex, iterative, and primarily social process. Widmann and Mulder (2018) state that interaction, especially reflexivity, is related positively to innovative work behavior.

We found that OAC influenced science teachers' innovative work behavior. It can be assumed that teachers' innovative work behavior leads to innovation within the organization. Ming and Ying (2010) analyzed the effects of affective commitment on technological innovation and administrative innovation. They applied SEM to analyze whether and how affective commitment influenced organizational innovation. Ming and Ying (2010) have found that affective commitment had a directly significant and positive effect on both technological innovation and administrative innovation. A lot of research remains to be done to study the simultaneous influence of the three forms of organizational commitments on science teachers' innovative work behavior abilities.

### 3.3.5. Conclusions

In this chapter, we strove to provide the most reliable portrait possible of OAC's influence on science teachers' innovative work behavior. On the basis of our analyses of the TIMSS 2015 data using CFA and SEM techniques, we found that OAC had a statistically significant influence on science teachers' innovative work behavior.

The CFA analysis revealed that organizational/affective commitment is described by a set of TIMSS 2015 questionnaire variables: I am content with my profession as a teacher; I am satisfied with being a teacher at this school; I find my work full of meaning and purpose; I am enthusiastic about my job; My work inspires me; I am proud of the work I do; I am going to continue teaching for as long as I can.

SEM disclosed varying levels of influence of OAC on science teachers' innovative work behavior abilities. The OAC had the biggest influence on science teachers' ability to share innovative ideas ($\beta = 0.419$, $p < 0.05$) and the lowest influence on applying new ideas in education ($\beta = 0.331$, $p < 0.05$).

## 3.4. The role of self-confidence in teaching science in the innovative work behavior of science teachers

> *The section analyzes the role of personal cognitive factors in the innovative work behavior of science teachers of Lithuania. We use the SEM in order to determine the influence of science teachers' self-confidence in teaching science on their innovative work behavior.*

### 3.4.1. Introduction

Educational innovation plays an important role in society. "For an individual, a nation, and humankind to survive and progress, innovation and evolution are essential. Innovations in education are of particular importance because education plays a crucial role in creating a sustainable future" (Serdyukov, 2017, p. 5). Realizing educational innovation requires understanding the meaning of innovation. Innovation is understood as "the successful introduction of a new thing or method" (Brewer & Tierney, 2012, p. 15). This means innovation have two components: a new idea and the change that results from the adoption that new idea. (Levitt, 2002).

Scholars distinguish a number of steps involved in the innovation process. Serdyukov (2017) states that innovation requires three major steps: an idea, its implementation, and the outcome (Serdyukov, 2017,). Scott and Burce (1994) also proposed three dimensions of innovation behavior: idea generation, idea promotion, and idea realization. According to Rogers (2003), the innovation-decision process has five steps: knowledge, persuasion, decision, implementation, and confirmation. De Jong and Hartog (2010) distinguish four steps in the innovation process: idea generation, idea exploration, idea championing, and idea implementation. Kanter (1988) distinguished two steps: idea generation and idea realization. In conclusion, there is no one-size-fits-all approach to determine and finalize the steps involved in the innovation process.

At different stages of innovations different innovative work behaviors occur. De Jong and Hartog (2010) cite Farr and Ford (1990) and define innovative work behavior as "an individual's behaviour that aims to achieve the initiation and intentional introduction (within a work role, group or organization) of new and useful ideas, processes, products or procedures" (p. 24). De Jong and Hartog (2010) identified four forms of innovative work behavior: idea generation, idea exploration, idea championing, and idea implementation. Janssen (2003) described a three-stage process—(a) intentional idea generation, (b) idea promotion, and (c) idea realization—within a work role, work group, or organization, in order to benefit role performance, the group, or the organization.

Innovations in education depend on teachers' innovative work behavior: "innovation behavior is the key to organizational innovation and also a crucial element of organizational innovative development. Therefore, innovation behavior involves learners generating innovative ideas and problem solutions during innovation activities through which they demonstrate the ability to practice innovation" (Chang, 2018, p. 18). It is important to uncover the factors that determine innovative work behavior.

Thurlings and coauthors (2015) performed a large-scale research in order to find which factors affect the innovative work behavior of teachers and found that there are three types of factors that influence teachers' innovative work behavior: demographic, individual, and organizational. According to this classification, self-efficacy belongs to the individual factor. Other scholars (Gong et al., 2009; Chang & Yang, 2012; Chang, 2018) revealed that creative self-efficacy effectively predicts individual innovation behavior and performance.

Self-efficacy is a structural component of self-confidence. Self-confidence has practical relevance to education (Sheldrake, 2016). Self-confidence of school students is analyzed by different aspects: self-confidence and interest in particular subjects (Viljaranta, Tolvanen, Aunola, & Nurmi, 2014); self-confidence and students attainment (Zell & Krizan, 2014); self-confidence and subjective values (Eccles, 2009; Sheldrake, 2016). Science teachers' confidence is analyzed from various points of view: comparative analysis of pre-service and in-service teachers' confidence (Megan, 2016); self-confidence and science knowledge (Appleton, 1992, Wendy Harte & Paul Reitano, 2015; Malandrakis, 2018); factors of primary school teachers' confidence (Yates, 1990); communication and confidence (Train & Miyamoto, 2017); preservice teacher's anxiety about teaching science, including low self-efficacy (Yürük, 2011); and engaging preservice teachers in various professional development opportunities to build confidence (Kenny, 2010; DeCoito, 2006); teachers' and students' confidence in learning mathematical concepts (Barrow et al., 2018). Self-confidence has been explored in many ways, including in relation to both students and teachers. However, there is a lack of research on self-confidence and innovative work behavior. Our study aims to fill this gap in research.

The concept of self-confidence is complex and joins two dimensions: "self-concept" and "self-efficacy." Sheldrake (2016) cites Bong and Skaalvik (2003) and states that within educational research, self-confidence has often been conceptualized and measured as "self-concept" and "self-efficacy" beliefs, usually specific to particular academic subjects. According to Sheldrake

(2016), "Self-concept broadly considers someone's beliefs about their abilities, integrating historical experiences (such as receiving particular grades or accomplishing difficult work) and current evaluative or interpretative beliefs (such as whether the student is 'doing well' or is 'good' at the subject). Alternately, self-efficacy considers someone's evaluative beliefs about their future capacities, such as their confidence in being able to gain a particular examination grade or to successfully accomplish a particular type of exercise" (p. 51). The TIMSS 2015 questions about science teachers' confidence are oriented in teachers' beliefs about their abilities, integrating historical experience and current beliefs (Tab. 3.4.1). Consequently, we explored the influence of science teachers' self-confidence on their innovative work behavior by the self-concept aspect. It is advisable to conduct future research with an emphasis not only on the self-concept of science teachers but also on their self-efficacy in investigating teachers' innovative work behavior.

The aim of this study is to reveal the influence of science teachers' self-confidence in teaching science on innovative work behavior in science classroom on the basis of TIMSS 2015 data.

### 3.4.2. Method of the research

A secondary analysis of the TIMSS 2015 data was performed according to the theoretical model predicting the influence of science teachers' self-confidence in teaching science on their innovative work behavior. The TIMSS 2015 instrument for science teachers allowed for carrying out an empirical analysis of science teachers' confidence in teaching science. We used BTBS 17 question for our analysis, which was worded as follows: "In teaching science to this class, how would you characterize your confidence in doing the following?" The answers to this question revealed teachers' confidence in teaching activities in the science class (inspiring, explaining, adapting, helping, assessing, improving, making, developing, and using inquiry). We analyzed science teachers' activities in the classroom on the basis of both teaching strategies and teaching outcomes. Cognitive psychology highlights four strategies of knowledge construction: retrieval practice, spaced practice, feedback practice, and interleaving practice (Dunlosky et al., 2013; Tab. 3.4.1).

Tab. 3.4.1: Science teachers' confidence in teaching science: strategies and outcomes

| Code of question | Question | Strategy | Outcomes |
|---|---|---|---|
| BTBS 17A | Inspiring students to learn science | Retrieval practice | Motivation |
| BTBS 17B | Explaining science concepts or principles by doing science experiments | Retrieval practice | Achieving |
| BTBS 17C | Providing challenging tasks for the highest achieving students | Spaced practice | Highest achieving |
| BTBS 17D | Adapting my teaching to engage students' interest | Retrieval practice | Motivation |
| BTBS 17E | Helping students appreciate the value of learning science | Retrieval practice | Value |
| BTBS 17F | Assessing student comprehension of science | Feedback practice | Comprehension |
| BTBS 17G | Improving the understanding of struggling students | Spaced practice | Understanding |
| BTBS 17H | Making science relevant to students | Retrieval practice | Understanding |
| BTBS 17I | Developing students' higher-order thinking (HOT) skills | Retrieval practice, spaced practice, interleaving practice | Higher-order thinking skills |
| BTBS 17J | Teaching science using inquiry methods | Interleaving | Motivation, understanding, values of learning science |

The normality of the data for these questions was checked. A Kolmogorov–Smirnov test was performed (Tab. 3.4.2). The test results indicate that the data do not meet conditions of normality. Another method (skewness and kurtosis) was used to check the normality of the data. The values for asymmetry (skewness and kurtosis) between –2 and +2 are considered acceptable in order to prove normal univariate distribution (George & Mallery, 2010). Asymmetry coefficients indicate that data satisfy the condition of normality (Tab. 3.4.3).

Tab. 3.4.2: Normality of the data on science teachers' confidence in teaching science: Kolmogorov–Smirnov test

|  | BTBS 17A | BTBS 17B | BTBS 17C | BTBS 17D | BTBS 17E | BTBS 17F | BTBS 17G | BTBS 17H | BTBS 17I | BTBS 17J |
|---|---|---|---|---|---|---|---|---|---|---|
| Kolmogorov–Smirnov Z | 9.549 | 8.909 | 9.795 | 10.164 | 9.707 | 10.265 | 9.404 | 8.918 | 10.034 | 8.532 |
| Asymp. Sig. (2-tailed) | .000 | .000 | .000 | .000 | .000 | .000 | .000 | .000 | .000 | .000 |

Tab. 3.4.3: Normality of the data on science teachers' confidence in teaching science: asymmetry coefficients test

|  | BTBS 17A | BTBS 17B | BTBS 17C | BTBS 17D | BTBS 17E | BTBS 17F | BTBS 17G | BTBS 17H | BTBS 17I | BTBS 17J |
|---|---|---|---|---|---|---|---|---|---|---|
| Skewness | .221 | .371 | .478 | .113 | .201 | .187 | .204 | .358 | .269 | .130 |
| SE | .079 | .079 | .079 | .079 | .079 | .079 | .079 | .079 | .079 | .079 |
| Kurtosis | -.329 | -.438 | .273 | -.247 | -.596 | -.040 | -.423 | -.705 | -.043 | -.365 |
| SE | .158 | .158 | .158 | .158 | .158 | .158 | .158 | .158 | .159 | .158 |

*Conceptual framework of the influence of science teachers' self-confidence in teaching science on their innovative work behavior*

On the basis of our literature review, we created a conceptual framework of the influence of science teachers' self-confidence in teaching science on their innovative work behavior (Fig. 3.4.1). Science teachers' self-confidence in teaching science (STS) is a latent factor. The STS (an unobserved variable) is depicted graphically with ovals (Fig. 3.4.1).

The modeling of STS analysis is driven by theoretical relationships between the observed variables (BTBS 17A to BTBS 17J) and the unobserved variable. A CFA is conducted to minimize the difference between the estimated and observed matrices (Fig. 3.4.1).

# Science teacher's self-confidence in teaching science

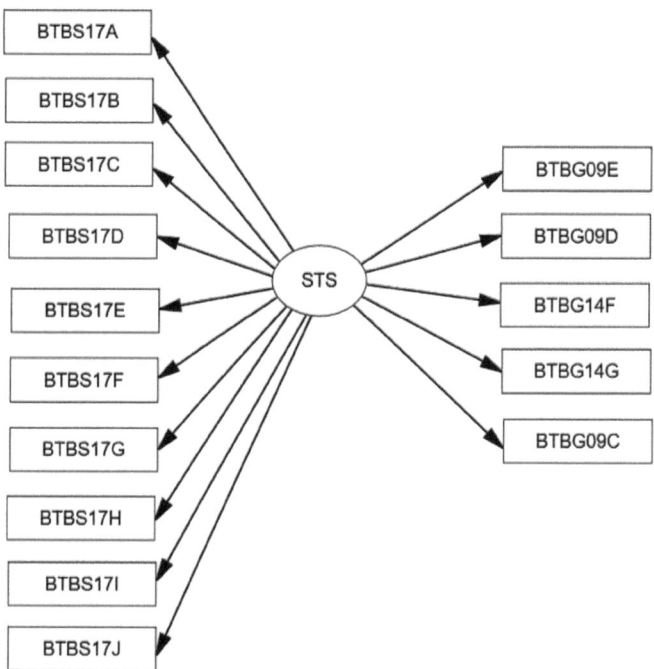

**Fig. 3.4.1:** Conceptual framework of science teachers' self-confidence in teaching science and innovative work behavior

We used the SEM to ascertain the influence of science teachers' self-confidence in teaching science on their innovative work behavior. Schreiber et al. (2006) state that "SEM, in comparison with CFA, extends the possibility of relationships among the latent variables and encompasses two components: (a) a measurement model (essentially the CFA) and (b) a structural model" (p. 325). We used the SEM to test five hypotheses (Fig. 3.4.1):

$H_1$. Science teachers' self-confidence in teaching science affects science teachers' abilities to work together trying out new ideas.

$H_2$. Science teachers' self-confidence in teaching science affects science teachers' abilities to visit another classroom to learn more about teaching.

$H_3$. Science teachers' self-confidence in teaching science affects science teachers' abilities to ask students to decide their own problem-solving procedures.

H$_4$. Science teachers' self-confidence in teaching science affects science teachers' abilities to encourage students to express their ideas in class.

H$_5$. Science teachers' self-confidence in teaching science affects science teachers' abilities to share what they have learned about their teaching experiences.

### 3.4.3. Results of the research

*Results of the exploratory factor analysis (EFA) of science teachers' self-confidence in teaching science*

An EFA was used for investigating the relationships between the variables of a complex concept such as science teachers' self-confidence in teaching science. The EFA is used to reduce data (BTBS 17A to BTBS 17J) to a smaller set of summary variables and to explore the underlying theoretical structure of science teachers' self-confidence in teaching science.

Our EFA of science teachers' self-confidence in teaching science comprised the following steps: reliable measurements, correlation matrix, principal components analysis (PCA), factor rotation, and the use and interpretation of results (Rietveld & Van Hout, 1993). The application of exploratory factor analysis have taken into account variables that can be measured at a range level, normally distributed (Field, 2000). The skewness (from −1 to +1) and kurtosis (from −1 to +1) of variable from the questions group M1 were well within the tolerable range for assuming a normal distribution. The coefficients of asymmetry (skewness and kurtosis) confirmed data normality (Tab. 3.4.2.).

The KMO test was used for checking sampling adequacy. It was found that KMO = 0.961 > 0.05 for observed variables of the group of questions addressing science teachers' self-confidence in teaching science. The intercorrelations between the studied variables were checked. Intercorrelation was checked by using Bartlett's test: $\chi^2$ (6) = 7,306.292, $p < 0.05$. It means that data about science teachers' self-confidence in teaching science may be used for factor analysis.

Factorability of research data was assured by anti-image correlation. All elements on the diagonal (MSA) of anti-image correlation matrix should be greater than 0.5 if the sample is adequate (Field, 2000, p. 446). All variables are suitable for factor analysis (Tab. 3.4.4).

**Tab. 3.4.4:** Anti-image correlation matrix of questions about science teachers' self-confidence in teaching science

| Question code | BTBS 17A | BTBS 17B | BTBS 17C | BTBS 17D | BTBS 17E | BTBS 17F | BTBS 17G | BTBS 17H | BTBS 17I | BTBS 17J |
|---|---|---|---|---|---|---|---|---|---|---|
| BTBS 17A | .966[a] | | | | | | | | | |
| BTBS 17B | -.185 | .958[a] | | | | | | | | |
| BTBS 17C | -.078 | -.174 | .961[a] | | | | | | | |
| BTBS 17D | -.148 | -.116 | -.176 | .968[a] | | | | | | |
| BTBS 17E | -.223 | -.006 | .002 | -.141 | .959[a] | | | | | |
| BTBS 17F | -.073 | -.093 | -.074 | -.117 | -.198 | .962[a] | | | | |
| BTBS 17G | -.065 | -.051 | -.093 | -.079 | -.037 | -.257 | .965[a] | | | |
| BTBS 17H | -.096 | -.082 | -.003 | -.150 | -.195 | -.121 | -.195 | .965[a] | | |
| BTBS 17I | -.037 | .094 | -.222 | -.066 | -.070 | -.046 | -.072 | -.109 | .950[a] | |
| BTBS 17J | -.054 | -.218 | -.154 | -.050 | .004 | -.068 | -.059 | -.056 | -.298 | .954[a] |

a. Measures of sampling adequacy (MSA).

Initial eigenvalues of the PCA indicated that there was only one factor that explained the variance (Tab. 3.4.5). This factor (Tab. 3.4.5) corresponds to the Guttman–Kaiser rule because its eigenvalues > 1 and explains 67.875% of variance. It means that only one factor is appropriate for data. Only one component was extracted. It means that the solution cannot be rotated.

**Tab. 3.4.5:** Eigenvalues percentage of variance and cumulative percentages for the factor science teachers' self-confidence in teaching science

| Factor | Eigenvalues | % of Variance | Cumulative % |
|---|---|---|---|
| 1 | 6.787 | 67.875 | 67.875 |

Extraction method: Principal component analysis.

As the last step of the EFA, factor loadings were determined (Tab. 3.4.6). All variables *have* a strong association with the factor of science teachers' self-confidence in teaching science (Tab. 3.4.6).

**Tab. 3.4.6:** Observed variables and loadings of the factor science teachers' self-confidence in teaching science

| Observed variable | Factor loading |
|---|---|
| **Inspiring** students to learn science | .854 |
| **Explaining** science concepts or principles by doing science experiments | .851 |
| **Providing** challenging tasks for the highest achieving students | .843 |
| **Adapting** my teaching to engage students' interest | .830 |
| **Helping** students appreciate the value of learning science | .827 |
| **Assessing** student comprehension of science | .824 |
| **Improving** the understanding of struggling students | .813 |
| **Making** science relevant to students | .809 |
| **Developing** students' higher-order thinking (HOT) skills | .797 |
| Teaching science **using inquiry** methods | .786 |

## *Results of Confirmatory factor analysis (CFA)*

The latent construct, science teachers' self-confidence in teaching science, was examined using the CFA. Initially, the fit of the data to the model was checked. Bentler (1990) suggested that the measurement model fits the data well if:

- $\chi^2/df$ is less than 5, describes the distance between model and data, but depends on sample size;
- The RMSEA is less than 0.08, describes how much error or unexplained variance remains after fitting the model;
- NFI is larger than 0.80;
- IFI exceeds 0.90;
- TLI exceeds 0.90, describes the "power" of model compared to "the situation without the model";
- CFI exceeds 0.90, describes the "power" of model compared to "the situation without the model" (Bentler, 1990).

The fitness of items of the latent factor, that is, science teachers' self-confidence in teaching science, revealed a sufficient fit and confirmed ten questionnaires' structure (Tab. 3.4.7). The CFA confirmed that the structural model fit the data well (Fig. 3.4.1).

**Tab. 3.4.7:** The fitness of items of the latent factor science teachers' self-confidence in teaching science

|  | Absolute fit index | | | Relative fit index | | |
|---|---|---|---|---|---|---|
|  | $\chi^2/df$ | RMSEA | GFI | IFI | TLI | CFI |
| Assumed model | 2.212 | .036 | .977 | .991 | .987 | .990 |
| Acceptance value | 1–5 | <.08 | >.80 | >.90 | >.90 | >.90 |

Unstandardized coefficients for observed variables and latent factor Confidence in teaching science (CTS) were deducted (Tab. 3.4.8). The unstandardized beta ($B$) value represents the predictor (observed) variable and the dependent (latent) variable. The unstandardized beta ($B$) value represents the influence of the predictor variable on the dependent variable. For the variable *inspiring students to learn science*, this would mean that for every one-unit increase in the variable *inspiring students to learn science*, the dependent variable (STS) increases by 1.00 unit. Also similarly, for the variable *explaining science concepts or principles by doing science experiments*, for every one-unit increase in the variable *explaining science concepts or principles by doing science experiments*, the dependent variable (STS) increases by 1.017 units.

**Tab. 3.4.8:** Standardized and unstandardized coefficients of the latent variable science teachers' self-confidence in teaching science: the latent construct "science teachers' confidence in teaching science" (CTS)

| Code of observed variable | Observed variable | B | β | SE | p label |
|---|---|---|---|---|---|
| BTBS 17A | **Inspiring** students to learn science | 1.000 | .817 |  |  |
| BTBS 17B | **Explaining** science concepts or principles by doing science experiments | 1.017 | .777 | .037 | <.001 |
| BTBS 17C | **Providing** challenging tasks for the highest achieving students | 1.039 | .793 | .033 | <.001 |
| BTBS 17D | **Adapting** my teaching to engage students' interest | 1.042 | .845 | .034 | <.001 |
| BTBS 17E | **Helping** students appreciate the value of learning science | .997 | .774 | .034 | <.001 |
| BTBS 17F | **Assessing** student comprehension of science | 1.018 | .825 | .036 | <.001 |
| BTBS 17G | **Improving** the understanding of struggling students | 1.015 | .791 | .034 | <.001 |
| BTBS 17H | **Making** science relevant to students | 1.002 | .818 | .040 | <.001 |
| BTBS 17I | **Developing** students' higher-order thinking skills | 1.034 | .743 | .039 | <.001 |
| BTBS 17J | Teaching science **using inquiry** methods | 1.046 | .768 | .037 | <.001 |

The use of nquiry methods mostly influence science teachers' self-confidence for teaching science because the value of the unstandardized beta is the highest (Tab. 3.4.8).

SE denotes the standard error for the unstandardized beta (*SEB*; Tab. 3.4.8). This value is similar to the standard deviation for a mean. The larger the value of SE, the more spread out the points are from the regression line and the less likely that significance will be found.

The standardized beta ($\beta$) works very similarly to a correlation coefficient. With this, we can compare the observed variables to see which one of them had the strongest relationship with the dependent variable (since all of them are on the 0–1 scale). Science teachers' self-confidence in *adapting the teaching to engage students' interest* has the strongest relationship with the latent variable *science teachers' confidence for teaching science* ($\beta$ = 0.845; Tab. 3.4.8). Science teachers' self-confidence in *adapting the teaching to engage students' interest* corresponds to retrieval practice (Tab. 3.4.1). The outcome of *adapting the teaching to engage students' interest* is students' intrinsic motivation for learning science.

The lowest relationship with latent variable (*science teachers' self-confidence in teaching science*) was observed in the case of the variable *developing students' higher-order thinking skills* ($\beta$ = 0.743; Tab. 3.4.8). The development of higher-order thinking (HOT) skills is a complex and complicated educational process where students would have to understand the facts, infer them, and connect them to other concepts.

"HOT takes thinking to higher levels than restating the facts and requires students to do something with the facts—understand them, infer from them, connect them to other facts and concepts, categorize them, manipulate them, put them together in new or novel ways, and apply them as we seek new solutions to new problems" (Thomas & Thorne, 2009). It means that development of students' HOT skills requires the teacher's ability to apply various teaching techniques. Therefore, it is clear that teachers have the least self-confidence for *developing students' higher-order thinking skills* ($\beta$ = 0.743; Tab. 3.3.3.7).

*Results of structural equation modeling (SEM)*

According to the theoretical model of innovative work behavior abilities, the main innovative work behavior abilities of the science teachers include the following: generation of new ideas in education, development of new ideas in education, applying new ideas in education, promotion of new ideas in education, and the modification and sharing of new ideas (Janssen, 2003).

The SEM framework (see Fig. 3.4.1) of science teachers' innovative work behavior abilities was built on the basis of the theoretical model of Janssen (2003). "The structural model displays the interrelations among latent constructs and observable variables in the proposed model as a succession of structural equations—akin to running several regression equations" (Schreiber et al., 2006, p. 325). The structural model (Fig. 3.4.2) displays the interrelations between the exogenous (unobserved) variable (science teachers' self-confidence in teaching science [STS]) and endogenous (observed) variables (science teachers' innovative work behavior abilities) through a succession of structural equations. The direct effect of the exogenous variable (STS) on an exogenous variable (science teachers' innovative work behavior abilities; Fig. 3.4.2) was investigated. The SEM allowed us to test five hypotheses ($H_1$ to $H_5$) regarding how constructs are theoretically linked to significant relationships in direct effect. Tab. 3.4.9 demonstrates the results of our hypotheses testing. All pathways are statistically significant (Tab. 3.4.9).

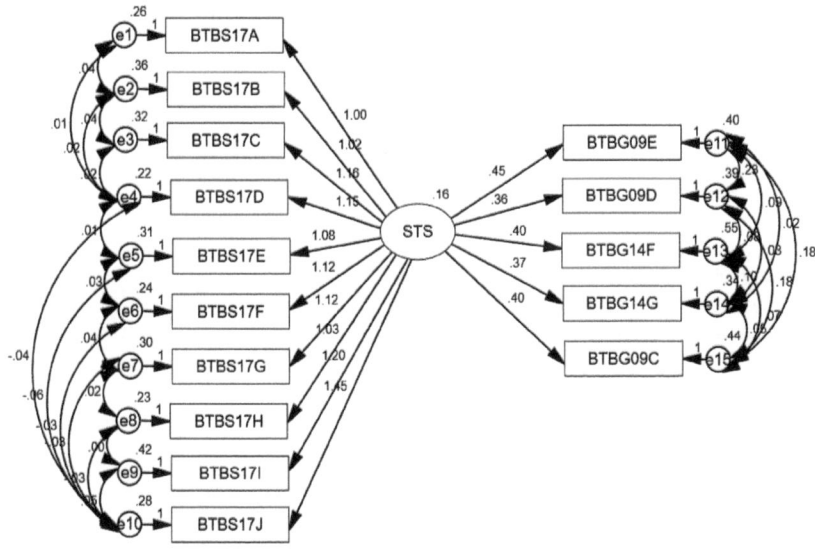

**Fig. 3.4.2:** SEM results of science teachers' innovative work behavior abilities and science teachers' confidence in teaching science: unstandardized data

**Tab. 3.4.9:** The influence of science teachers' self-confidence in teaching science on science teachers' innovative work behavior abilities: path coefficients and statistical significance

| Hypothesis | Paths | Path coefficients ($\beta$) | $p$ | $R^2$ | Results |
|---|---|---|---|---|---|
| $H_1$. Science teachers' self-confidence in learning science affects science teachers' abilities to work together in trying out new ideas | Self-confidence in teaching science (STS) → Work together to **try out new ideas** (BTBG 09E) | .376 | <.001 | .142 | Support |
| $H_2$. Science teachers' self-confidence in learning science affects science teachers' abilities to visit another classroom to learn more about teaching | Self-confidence in teaching science (STS) → Visit another classroom to **learn more** about teaching (BTBG 09D) | .386 | <.001 | .149 | Support |
| $H_5$. Science teachers' self-confidence in learning science affects science teachers' abilities to share what they have learned from their teaching experiences | Self-confidence in teaching science (STS) → **Share what I have** learned about my teaching experiences (BTBG 09C) | .349 | <.001 | .122 | Support |
| $H_3$. Science teachers' self-confidence in learning science affects science teachers' abilities to ask students to decide their own problem-solving procedures | Self-confidence in teaching science (STS) → **Ask students** to decide their own problem-solving procedures (BTBG 14F) | .524 | <.001 | .275 | Support |
| $H_4$. Science teachers' self-confidence in learning science affects science teachers' abilities to encourage students to express their ideas in class | Self-confidence in teaching science (STS) → **Encourage students** to express their ideas in class (BTBG 14G) | .557 | <.001 | .310 | Support |

We performed hypothesis ($H_1$ to $H_5$) testing results according to the results of EFA of science teachers' innovative work behavior abilities. The EFA of science teachers' innovative work behavior revealed two factors: idea generating and sharing; idea applying (Tab. 3.5.7).

Three variables fall under the first factor: visit another classroom to learn more about teaching ($H_2$), work together to try out new ideas ($H_1$), and share what they have learned about my teaching experiences ($H_5$). Two variables fall under the second factor: encourage students to express their ideas in class ($H_4$) and ask students to decide their own problem-solving procedures ($H_3$).

We also performed hypotheses testing with $R^2$. $R^2$ is a statistical coefficient that represents the percentage of the response variation that is explained by a linear model. In other words, $R^2$ measures how close the data are to the fitted regression line. It is also known as the coefficient of determination. The higher the $R^2$, the better the model fits the data. The data of $R^2$ show that the model has a better fit to data of EFA of the factor Idea Applying.

The coefficient of determination ($R^2$) value for the variable *science teachers' abilities to ask students to decide their own problem-solving procedures* was 0.275. This means that 27.5% of the teachers' ability to ask students to decide their own problem-solving procedures was influenced by science teachers' self-confidence in teaching science. The remaining changes are influenced by other factors. Science teachers' ability to encourage students to express their ideas in the class was 0.310 (Tab. 3.4.9). This is the greatest value of $R^2$. This means that 31.0% of the teachers' ability to encourage students to express their ideas in class was influenced by science teachers' self-confidence in teaching science. The remaining 69.0% changes are influenced by other factors. It should be noted that the values of $R^2$ are smaller for the first EFA factor—Idea generating and sharing (Tab. 3.4.9).

SEM results show that path coefficients for the first factor (Idea generating and sharing) and the second factor (Idea applying) differ almost twice (Tab. 3.4.9). The path coefficients are lower in the first factor (Idea generating and sharing) group (Tab. 3.4.9).

Analysis of path coefficients in the second factor (Idea applying) confirmed that path coefficients are different in this case as well. Our data indicate that science teachers' innovative ability to encourage students to express their ideas in class (EIC) is directly affected by STS. We regress science teachers' innovative ability to encourage students to express their ideas in class on STS variables and obtain $R^2 = 0.310$, $\beta_{EIC\text{-}STS} = 0.557$. We also regress science teachers' innovative ability to ask students to decide their own problem-solving procedures (APS) on STS variables—$R^2 = 0.275$, $\beta_{APS\text{-}STS} = 0.524$. It can be argued that science teachers' self-confidence in teaching science has more influence on science teachers' innovative ability to encourage students to express their ideas in class (EIC) and their ability to ask students to decide their own problem-solving procedures (APS).

### 3.4.4. Discussion

In the present study, we analyzed the influence of science teachers' self-confidence in teaching science on their innovative work behavior abilities. We investigated teachers' innovative work behavior by carrying out a secondary analysis of the TIMSS 2015 data. The theoretical background of our investigation was based on diffusion theory (Rogers, 2003). We used the EFA for investigating the variable relationships of a complex concept such as the science teachers' innovative work behavior abilities on the basis of TIMSS 2015 questionnaire. The EFA revealed two factors: idea generating and sharing; idea applying. The SEM displayed that path coefficients are lower in idea generating and the sharing factor (Tab. 3.4.9). This means that science teachers' self-confidence in teaching science has lesser influence on their idea generating and sharing abilities and greater influence on their ability to apply ideas. Serdyukov (2017) posits that "innovation requires three major steps: an idea, its implementation, and the outcome that results from the execution of the idea and produces a change" (p. 8). The SEM results show that science teachers' self-confidence in teaching science exerts more influence on their innovative work behavior at the second stage—idea implementation.

This study investigated science teachers' teaching activity and self-confidence in teaching science in the light of cognitive psychology (Tab. 3.4.1). Scholars state that teaching is not just about facts, but that it is a process of coming to understand the world (Agarval et al., 2012; Agarval & Roediger, 2018). The process of world understanding is based on knowledge construction in one form or another. According to cognitive psychologists, the learning process follows four strategies: retrieval practice, spaced practice, feedback practice, and interleaving practice (Dunlosky et al., 2013). Retrieval practice boosts learning by pulling information out of students' heads, rather than cramming information into their heads. Feedback practice serves for revealing to students what they know and don't know. Feedback practice increases students' understanding about their own learning progress. Spaced practice promotes returning to learning content every so often to help students consolidate and refresh their knowledge. Interleaving practice boosts learning by encouraging connections between and among closely related topics. Teachers implement these strategies through various activities: inspiring students to learn science, explaining science concepts or principles by doing science experiments, providing challenging tasks for the highest achieving students, engaging students' interest, helping students appreciate the value of learning science, assessing students' comprehension of science, improving the understanding of struggling students, making science relevant to students, developing students' higher-order thinking skills, and teaching science using inquiry methods (Tab. 3.3.1). The effectiveness of these activities in science classroom depends on both personal and organizational factors (Deary, Strand, Smith, & Fernandes, 2007; Rohde & Thompson, 2007; Meyer et al., 2019, p. 58). The CFA revealed the highest pathway coefficients in science teachers' self-confidence in teaching science occurred with different strategies: spaced practice, retrieval practice, and interleaving practice. This study also recommends future studies to investigate science teachers' self-confidence in teaching science in the light of cognitive psychology.

We analyzed the influence of science teachers' self-confidence in teaching science on their innovative work behavior abilities at the individual level. Scholars state that the field of innovation ranges from the organizational level to the individual level (Axtell et al., 2000). At the individual level, innovation occurs in the implementation of small-scale ideas that are related to improvements in daily work processes and activities (Axtell et al., 2000). In the present study, teachers' innovative work behavior is studied at the individual level as teachers are the ones who primarily contribute to small-scale innovations in the domain of their work roles and initiate the process of innovation in their teaching. It would make sense to repeat our study at the next level, that of the organization.

Our research was constrained by limitations. The measurement of innovative work behavior is challenging. "Both scientists and practitioners emphasize the importance of innovative work behaviour (IWB) of individual employees for organizational success, but the measurement of IWB is still at an evolutionary stage" (De Jong & Hartog, 2010). We didn't use a special questionnaire to investigate the innovative work behavior of science teachers. We conducted a secondary analysis of science teachers' innovative work behavior using the TIMSS 2015 questionnaire. We used only one sample—the sample of science teachers from Lithuania. It would be useful to repeat the study with the database of other countries that participated in TIMSS 2015.

### 3.4.5. Conclusions

Science teachers implement learning strategies through various activities. Statistical analysis of various activities (inspiring students to learn science, explaining science concepts or principles by doing science experiments, providing challenging tasks for the highest achieving students, engaging students' interest, helping students appreciate the value of learning science, assessing students' comprehension of science, improving the understanding of struggling students, making science relevant to students, developing students' higher-order thinking skills, teaching science using inquiry methods) revealed that the usage of inquiry methods maximum influence on science teachers' self- confidence in teaching science. Science teachers' activity of helping students appreciate the value of learning science had the least influence on their confidence in teaching science.

The present study found that science teachers' self-confidence in teaching science had a statistically significant positive influence on the innovative work behavior of science teachers. The EFA revealed two groups of science teachers' innovative work behavior abilities: Idea generating and sharing; Idea applying.

SEM results confirmed the influence of science teachers' self-confidence in teaching science on their innovative work behavior abilities. Self-confidence of science teachers in teaching science had more influence on idea applying abilities but less influence on idea generating and sharing abilities.

Analysis of path coefficients in the second factor (idea applying) confirmed that science teachers' innovative ability to encourage students to express their ideas in class and innovative ability to ask students to decide their own problem-solving procedures are directly and positively affected by science teachers' self-confidence in teaching science.

## 3.5. The influence of organization leadership support on science teachers' innovative work behavior

> *In this section we analyze the role of environmental factor—organizational leadership support—using confirmatory factor analysis and structural equation modeling. The results of these analyses are interpreted according to Social Cognitive theory.*

### 3.5.1. Introduction

There has been a lot of research on organizational innovation by the aspect of members' leadership in an organization. A lot of research has sought to identify factors that can stimulate organizational innovation (Tsai & Tseng, 2010). Scholars revealed that leadership is one of the key factors that affect organizational innovation (Gumusluoglu & Ilsev, 2009b; Makri & Scandura, 2010; Prasad & Junni, 2016; Mokhber et al., 2015; Mokhber et al., 2018). The role of leadership of organization members in organizational innovation is widely analyzed in the context of transformational leadership (Jung, Chow, & Wu, 2003, 2008; Gumusluoglu & Ilsev, 2009a, 2009b; Prasad & Junni, 2016; Feng, Huang, & Zhang, 2016). Mokhber et al. (2018) cite Feng, Huang, and Zhang (2016) and state that "Transformational leaders are able to support organizational innovation by enhancing the motivation and ability of organizational members to be creative and innovative. Transformational leaders develop enthusiasm among organizational members to think out-of-the-box and be more creative and to develop new ideas and solutions concerning organizational structures, processes and practices" (p. 109).

Organizational leadership support on innovative work behavior depends on the nature of the organization.

The role of leadership in education is illustrated by the *wedding cake* metaphor (Frost, 2010). Frost (2010) describes four layers of a wedding cake: student learning, teacher learning, school learning, and system learning. The layers of the cake are united (bonded) by leadership. Defying leadership in the wedding cake model means that leadership in education can be seen like a human capacity that can be exercised by everyone: from students to teachers to the school system to parents. Frost (2008) states that teacher leadership is like a sleeping giant that must be awakened without delay. The wedding cake model of leadership at the school highlights the role of communication in distributing ideas between different layers: student, teacher, parents, and the school system. Frost (2008) states that teacher leadership is often associated with distributed leadership. "A distributed perspective also suggests that leadership activity at the level of the school, rather than at the level of an individual leader, is the appropriate unit for studying leadership practice" (Spillane et al., 2001, p. 27). Goddard et al. (2015) states that leadership influences the teachers' ability to work together. Moreover, leadership and teacher collaboration may contribute to school effectiveness and collective efficacy (Goddard et al., 2015).

A lot of research has been done on organizational innovation based on teachers' leadership in school. The relationship between teachers' leadership and innovation within an organization is twofold: On one hand, teacher's leadership determines the innovation processes within the organization, and, on the other, organizational innovation leads to innovative work behavior of teachers. For decades, scholars have been studying the significance of organizational innovation for employees' innovative work behavior (Jung et al., 2003; Gumusluoglu & Ilsev, 2009a; Mokhber et al., 2018). However, research in the reverse direction (organizational support for employee's leadership) is missing: "support for innovative work behaviors as an innovation climate in the organization received only limited attention as a moderator" (Mokhber et al., 2018, p. 113). More research is being done to investigate the significance of organization leadership support

(OLS) for teachers' innovative work behavior because "the stimulation of organizational innovation is highly dependent on an organization that encourages innovation" (Mokhber et al., 2018, p. 113). According to Oke, Munshi, and Walumbwa (2009), leadership support encourages innovative work behavior and supports innovative culture in the organization (Mitchell & Tarter, 2016).

Science teachers' innovative work behavior plays a key role in the quality of science education, particularly in improving STEM (science, technology, engineering, mathematics) education. McKay et al. (2018) describe the structures that have fostered teacher leadership and how those structures emerged through partnership and collaboration and how they influence innovation in science education. Innovative work behavior of teachers was analyzed by different aspects: innovation activity (Sirotin & Arhipova, 2015), an innovative opportunity (Uddin, Fan, & Das, 2017), and innovative thinking (Ilinykh & Udaltsova, 2016). "It is recommended to do a deeper study of effect of external factors, such as the organizational culture of the school, the current school management model, the style of leadership, the activities of the school leader to form teacher's innovative work behavior, etc." (Trapitsin et al., 2018).

This situation necessitates for a deeper look into the problem of OLS for teachers' innovative work behavior. The discussed situation highlights *the scientific problem* that is formulated as a question: How does OLS influence science teachers' innovative work behavior?

*The object of the research* is the science teachers' innovative work behavior.

The aim of this study is to reveal the influence of organizational leadership support on science teachers' innovative work behavior.

### 3.5.2. Method of the research

A new framework is proposed in this study to determine the existence of organizational leadership support for science teachers' innovative work behavior. The research methodology is based on the assumption that innovative work behaviors are closely related to organizational leadership support, and, therefore, teachers (leaders) are the ones who establish organizational goals, make decisions on adopting and applying new ways of doing job, and motivate employees (Tsai & Tseng, 2010). Support for innovative work behaviors in the organization may provide a better communicating atmosphere for employees and leaders that ultimately contributes to organizational innovation.

A secondary analysis of the TIMSS 2015 data was performed according to the theoretical model of OLS for science teachers' innovative work behavior. The TIMSS 2015 instrument for science teachers allowed for carrying out an empirical analysis of science teachers' organizational leadership support (BTBG 10 question; Tab. 3.5.1).

Tab. 3.5.1: The influence of organization leadership support on science teachers' innovative work behavior variables according to TIMSS 2015

| Code | Items |
|---|---|
| BTBG 06A | How would you characterize each of the following within your school? Teachers' understanding of the school's curricular goals |
| BTBG 06O | How would you characterize each of the following within your school? Collaboration between school leadership and teachers to plan instruction |
| BTBG 06P | How would you characterize each of the following within your school? Amount of instructional support provided to teachers by school leadership |
| BTBG 06Q | How would you characterize each of the following within your school? School leadership's support for teachers' professional development |

*The aim of the research* (to reveal the influence of organizational leadership support on science teachers' innovative work behavior abilities) is illustrated in Fig. 3.5.1.

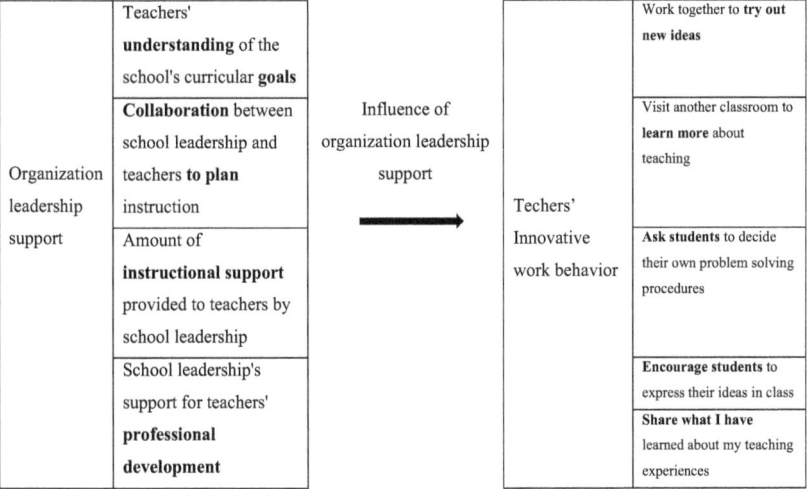

Fig. 3.5.1: The model of influence of organization leadership support on science teachers innovative work behavior

We checked data normality. The skewness and kurtosis were well within the tolerable range for assuming a normal distribution, and examination of the histograms suggested that the distributions looked approximately normal (Tab. 3.5.2). The values for skewness (asymmetry) and kurtosis between –2 and +2 are considered acceptable in order to prove normal univariate distribution (George & Mallery, 2010).

**Tab. 3.5.2a:** The skewness and kurtosis values of organization leadership support variables

|  | BTBG 06A | BTBG 06O | BTBG 06P | BTBG 06Q |
|---|---|---|---|---|
| Skewness | .292 | .710 | 2.340 | 2.518 |
| Std. Error of Skewness | .152 | .152 | .152 | .152 |
| Kurtosis | -.816 | 1.626 | 16.897 | 18.076 |
| Std. Error of Kurtosis | .302 | .302 | .302 | .302 |

The data of the variables BTBG 06Q and BTBG 06P were not corresponding to the requirement of normality (Tab. 3.5.2). We removed 19 questionnaires and rechecked the condition of data normality (Tab. 3.5.3). The rechecked data conformed to the normality requirements.

**Tab. 3.5.2b:** The skewness and kurtosis values of organization leadership support variables after removing extreme values

|  | BTBG 06A | BTBG 06O | BTBG 06P | BTBG 06Q |
|---|---|---|---|---|
| Skewness | .265 | .727 | .404 | .492 |
| Std. Error of Skewness | .154 | .154 | .154 | .154 |
| Kurtosis | -.851 | 1.671 | .562 | .094 |
| Std. Error of Kurtosis | .307 | .307 | .307 | .307 |

*Conceptual framework for the CFA carried out to study organization leadership support*

On the basis of the foregoing discussion in literature review, a conceptual framework for carrying out a CFA of organization leadership support was made (Fig. 3.5.2). The OLS framework comprises four latent variables: teachers' understanding of school's curricular goals, collaboration between school leadership and teachers to plan instruction, amount of instructional support provided to teachers by school leadership, and the school leadership's support for teachers' professional development (Fig. 3.5.2).

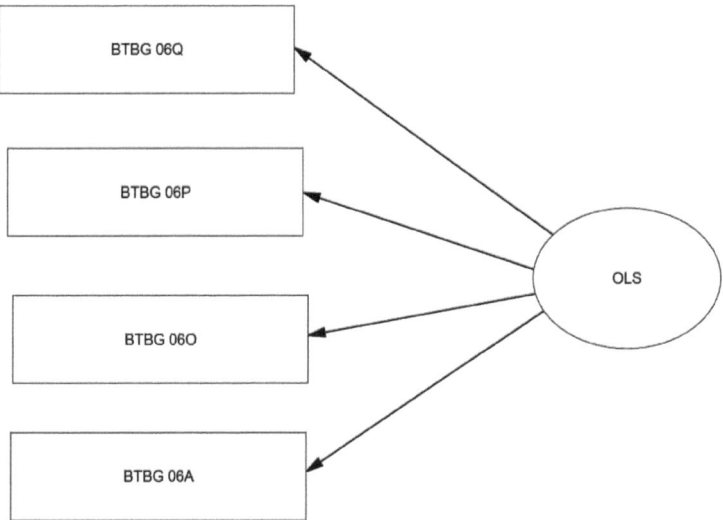

**Fig. 3.5.2:** Conceptual framework of school organization leadership support (OLS)

## *Conceptual framework of SEM utilized to study science teachers' innovative work behaviors*

The structural equation modeling (SEM) combines EFA and multiple regression analysis (Ullman, 2001). The SEM was deemed an appropriate technique for testing those relationships if the underlying structural patterns among all latent variables were informed theoretically. (Schreiber et al., 2006). The underlying structural pattern of OLS was created theoretically for developing a conceptual framework (Fig. 3.5.2).

The SEM framework (see Fig. 3.5.3) presents science teachers' innovative work behavior abilities based on a linear structural modeling. The structural model (Fig. 3.5.3) displays the interrelations among latent constructs (OLS) and observable variables (science teachers' innovative work behavior abilities) in the proposed model as a succession of structural equations. We are checking the direct effects among latent constructs as dictated by theory or empirically based suppositions. A direct effect (Fig. 3.5.3) represents the effect of an independent variable (exogenous) on a dependent variable (endogenous). In sum, "SEM allows researchers to test theoretical propositions regarding how constructs are theoretically linked and the directionality of significant relationships" (Schreiber et al., 2006, p. 326). We tested five hypotheses:

$H_1$. Organization leadership support affects science teachers' abilities to work together in trying out new ideas.

$H_2$. Organization leadership support affects science teachers' abilities to visit another classroom to learn more about teaching.

$H_3$. Organization leadership support affects science teachers' abilities to ask students to decide their own problem-solving procedures.

$H_4$. Organization leadership support affects science teachers' abilities to encourage students to express their ideas in class.

$H_5$. Organization leadership support affects science teachers' abilities to share what they have learned about their teaching experiences.

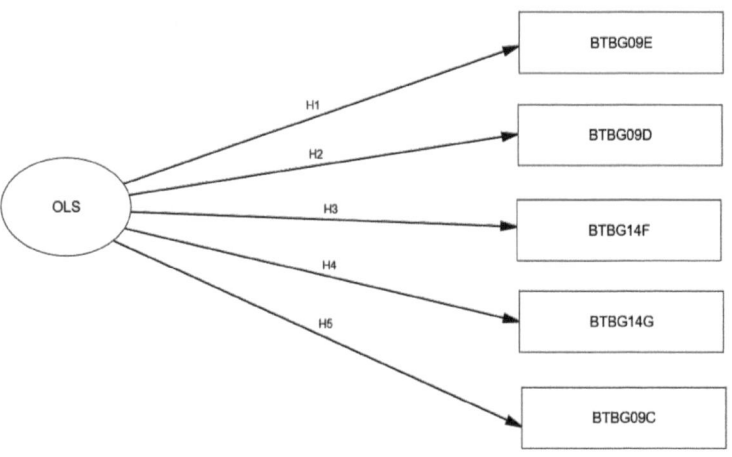

**Fig. 3.5.3:** SEM framework: conceptual framework of science teachers' innovative work behavior abilities and organization leadership support

### 3.5.3. Results of the research

*Results of the exploratory factor analysis (EFA) of organization leadership support (OLS)*

For investigating the variable relationships of complex concepts such as organization leadership, a exploratory factor analysis was used. Factor analysis allows for investigating concepts that are not easily measured directly by collapsing a large number of variables into a few interpretable underlying factors. The EFA was used for investigating the variable of organization leadership support. The EFA is used to reduce data to a smaller set of summary variables and explore the underlining theoretical structure of the phenomena to be studied (teachers' understanding of the school's curricular goals, collaboration between school leadership and teachers to plan instruction, amount of instructional support provided to teachers by school leadership, and school leadership's support for teachers' professional development). In our case, the observed variables can be associated with the latent variable—OLS.

In this study, the EFA of organization leadership support was carried out by Rietveld and Van Hout's (1993, p. 291) factor analysis diagram using the following steps: reliable measurements, correlation matrix, factor analysis versus PCA, number of factors to be retained, factor rotation, and the use and interpretation of results. In the application of factor analysis, range-level variable measurement has been taken into account (Field, 2000, p. 444). The skewness (from −1 to +1) and kurtosis (from −1 to +1) of variable from questions group M1 were well within the tolerable range for assuming a normal distribution.

Factorability was examined based on MSA. The KMO test was used for determining sample adequacy. The sample is adequate if the value of KMO is greater than 0.5. It was found that KMO = 0.779 for observed variables of the group of questions addressing OLS.

The starting point of factor analysis is a correlation matrix in which the intercorrelations between the studied variables are presented. The dimensionality of this matrix can be reduced by "looking for variables that correlate highly with a group of other variables but correlate very badly with variables outside of that group" (Field, 2000, p. 424). These variables with high intercorrelations could well measure one underlying variable, which is called a "factor." Intercorrelation was checked by using Bartlett's test: $\chi^2$ (6) = 372.240, $p < 0.05$. It means that data may be used for factor analysis.

Factorability of research data was assured by anti-image correlation. All elements on the diagonal (MSA) of anti-image correlation matrix should be greater than 0.5 if the sample is adequate (Field, 2000, p. 446). All variables are suitable for factor analysis (Tab. 3.5.3).

**Tab. 3.5.3:** Anti-image correlation matrix of questions about organization leadership support

|   | 1<br>Teachers' understanding of the school's curricular goals | 2<br>Collaboration between school leadership and teachers to plan instruction | 3<br>Amount of instructional support provided to teachers by school leadership | 4<br>School leadership's support for teachers' professional development |
|---|---|---|---|---|
| 1 | .0892 |  |  |  |
| 2 | -.168 | .808 |  |  |
| 3 | -.094 | -.353 | .736 |  |
| 4 | -.137 | -.261 | -.491 | 7.55 |

The PCA was used for investigating the variable of questions group addressing the organization leadership support. Initial eigenvalues indicated that a single factor explained 64.687% of the variance (Tab. 3.5.4). This factor (Tab. 3.5.4) corresponds to the Guttman–Kaiser rule because its eigenvalues > 1 explains 64.687% of variance. It means that only one factor is appropriate for data. Only one component was extracted. It means that the solution cannot be rotated.

**Tab. 3.5.4:** Eigenvalues percentage of variance and cumulative percentages for factor of the organization leadership support

| Factor | Eigenvalues | % of variance | Cumulative % |
|---|---|---|---|
| 1 | 2.587 | 64.687 | 64.687 |

Extraction method: Principal component analysis.

The last step in EFA was the interpretation of results based on factor loading. The relationship of each variable to the underlying factor is expressed by factor loading (Tab. 3.5.5). One could say that all variables *have* a strong association with the factor, the OLS.

**Tab. 3.5.5:** Observed variables and loadings of the factor organization leadership support

| Observed variable | Factor loading |
|---|---|
| Teachers' **understanding** of the school's curricular **goals** | .867 |
| **Collaboration** between school leadership and teachers **to plan** instruction | .859 |
| Amount of **instructional support** provided to teachers by school leadership | .835 |
| School leadership's support for teachers' **professional development** | .634 |

*Results of the exploratory factor analysis (EFA) of science teachers' innovative work behavior abilities*

A secondary analysis of the TIMSS 2015 data was performed according to the theoretical model of science teachers' innovative work behavior. The TIMSS 2015 instrument allowed us to carry out an empirical analysis of science teachers' innovative abilities: work together to try out new ideas, visit another classroom to learn more about teaching, ask students to decide their own problem-solving procedures, encourage students to express their ideas in class, and share what they have learned about my teaching experiences.

Five variables related to the factor of science teachers' innovative behavior were analyzed using the EFA with varimax (orthogonal) rotation. Factability of science teachers' innovative work behavior was examined based on by measures of sampling adequacy (MSA). The KMO test revealed sampling adequacy. It was found that KMO = 0.689 > 0.06 for observed variables. This value being more than 0.6 indicates the sampling is adequate and that remedial action shouldn't be taken. The intercorrelation was checked using Bartlett's test: $\chi^2$ (10) = 210.800, $p < 0.05$. All the elements on the diagonal (MSA) of anti-image correlation matrix were greater than 0.5 (Tab. 3.5.6.). Extraction communalities indicate that the variables chosen for this analysis were related to each other.

**Tab. 3.5.6:** Anti-image correlation matrix of questions about science teachers' innovative work behavior abilities

|   | Work together to try out new ideas | Visit another classroom to learn more about teaching | Ask students to decide their own problem-solving procedures | Encourage students to express their ideas in class | Share what they have learned from my teaching experiences |
|---|---|---|---|---|---|
|   | 1 | 2 | 3 | 4 | 5 |
| 1 | .661[a] | | | | |
| 2 | -.485 | .667[a] | | | |
| 3 | -.267 | -.227 | .760[a] | | |
| 4 | -.063 | -.083 | .035 | .684[a] | |
| 5 | -.061 | -.091 | .018 | -.187 | .703[a] |

a. Measures of sampling adequacy (MSA).

The PCA of science teachers' innovative behavior abilities yielded into two factors explaining a total of 64.542% of the variance for the entire set of variables (Tab. 3.5.7). Factor 1 was labelled *idea generating and sharing* due to the high loadings by the following items: work together to try out new ideas, visit another classroom to learn more about teaching, and share what they have learned about my teaching experiences. The first factor explained 42.243% of the variance. The second factor derived was labelled *idea applying*. This factor was labelled as such due to the high loadings by the following factors: Ask students to decide their own problem-solving procedures; encourage students to express their ideas in class. The variance explained by this factor was 22.259% (Tab. 3.5.7).

Tab. 3.5.7: The results of the principal component analysis (PCA) of science teachers' innovative behavior

| Factor | Eigenvalues | % of variance | Cumulative % | Rotation % of variance | Rotation % (cumulative) |
|---|---|---|---|---|---|
| 1 | 2.112 | 42.243 | 42.243 | 39.446 | 39.446 |
| 2 | 1.113 | 22.259 | 64.502 | 25.056 | 64.502 |

According to diffusion theory, the innovation-decision process involves five steps: (1) knowledge, (2) persuasion, (3) decision, (4) implementation, and (5) confirmation and sharing (Rogers, 2003). The number of steps can be reduced to idea generating (knowledge, persuasion, decision), idea applying (implementation), and idea sharing (confirmation and sharing). The EFA revealed that TIMSS 2015 variables corresponding to idea generating have high factor loadings (Tab. 3.5.8) with the factor idea generating and sharing like the variables corresponding to innovative idea implementation in science education.

Tab. 3.5.8: Factor analysis table of science teachers' innovative work behavior

| Items | Factor 1— idea generating and sharing | Factor 2— idea applying |
|---|---|---|
| Visit another classroom to learn more about teaching | .831 | |
| Work together to try out new ideas | .814 | |
| Share what they have learned from my teaching experiences | .779 | |
| Encourage students to express their ideas in class | | .779 |
| Ask students to decide their own problem-solving procedures | | .760 |
| Eigenvalue | 2.112 | 1.113 |
| % of total variance | 42.243 | 22.259 |
| Total variance | 64.542% | |

Extraction method: Principal component analysis.
Rotation Method: Varimax with Kaiser normalization.
a. Rotation converged in three iterations.

## CFA and SEM results: The influence of organization leadership support on science teachers' innovative work behavior

A CFA was performed in order to reveal the influence of organization leadership support (OLS) on science teachers' innovative work behavior. The CFA analysis is driven by the theoretical relationships among the observed variables (visit another classroom to learn more about teaching, work together to try out new ideas, share what they have learned about my teaching experiences, encourage students to express their ideas in class, and ask students to decide their own problem-solving procedures) and unobserved variable (OLS). We used a hypothesized model to estimate a science teachers' covariance matrix that is compared with the observed covariance matrix (Fig. 3.5.4). In performing the CFA we wanted to minimize the difference between the estimated and observed matrices.

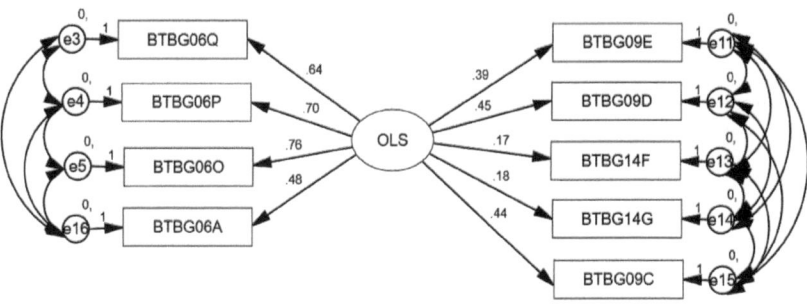

Fig. 3.5.4: SEM results of science teachers' innovative work behavior abilities and organization leadership support (OLS)

We performed an SEM analysis based on data from TIMSS 2015 with AMOS statistical package. We used the following goodness-of-fit indicators to assess a model (Fig. 3.5.4): NFI = 0.952 ≥ 0.95; NNFI (also known as TLI) = 0.986 ≥ 0.95; IFI = 0.991 ≥ 0.95; CFI = 0.991 ≥ 0.95; RMSEA = 0.029 < 0.06 to 0.08; and $\chi^2/df$ = 1.216 (confidence interval 1–5). The hypothesized model appears to be a good fit to the data.

Tab. 3.5.8 demonstrates the results of path analysis. Three statistically significant and two non–statistically significant pathways were identified. The EFA of science teachers' innovative abilities revealed two factors: idea generating and sharing; idea applying. Three statistically significant ways were identified in the first factor—idea generating and sharing. Two statistically insignificant pathways were detected in the factor idea applying.

It was found that OLS had a significant effect on science teachers' ability to learn more about teaching by visiting another classroom ($\beta = 0.454$, $p < 0.001$). Therefore, hypothesis H2 was supported. The OLS affects science teachers' ability to work together in trying out new ideas ($\beta = 0.391$, $p < 0.001$); OLS affects science teachers' abilities to share what they have learned ($\beta = 0.439$, $p < 0.001$). It means that hypotheses H1 and H3 were also supported (Tab. 3.4.4.8). However, OLS did not have a significant influence on science teachers' ability to apply new ideas: encouraging students to express their ideas in class ($\beta = 0.066$, $p > 0.05$); asking students to decide their own problem-solving procedures ($\beta = -0.175$, $p > 0.05$) (Tab. 3.4.4.9).

Tab. 3.5.9: The influence of organization leadership support on science teachers' innovative work behavior abilities: path coefficients and statistical significance

| Hypothesis | Paths | Path coefficients | $p$ | $R^2$ | Results |
|---|---|---|---|---|---|
| $H_1$. Organization leadership support affects science teachers' abilities to work together in trying out new ideas | Organization leadership support (OLS) → Work together to **try out new ideas** (**BTBG 09E**) | .391 | <.001 | .153 | Support |
| $H_2$. Organization leadership support affects science teachers' abilities to visit another classroom to learn more about teaching | Organization leadership support (OLS) → Visit another classroom to **learn more** about teaching (**BTBG 09D**) | .454 | <.001 | .206 | Support |
| $H_3$. Organization leadership support affects science teachers' abilities to share what they have learned | Organization leadership support (OLS) → **Share what I have** learned about my teaching experiences (**BTBG 09C**) | .439 | <.001 | .193 | Support |
| $H_4$. Organization leadership support affects science teachers' abilities to encourage students to express their ideas in class | Organization leadership support (OLS) → **Ask students** to decide their own problem-solving procedures (**BTBG 14F**) | .066 | .356 | .004 | Against |

Tab. 3.5.9: Continued

| Hypothesis | Paths | Path coefficients | p | $R^2$ | Results |
|---|---|---|---|---|---|
| $H_5$. Organization leadership support affects science teachers' abilities to ask students to decide their own problem-solving procedures about their teaching experiences | Organization leadership support (OLS) → **Encourage students** to express their ideas in class (**BTBG 14G**) | .175 | .015 | .031 | Against |

## 3.5.4. Discussion

The present research considered the processes activated in science teachers' innovative work behavior. We have shown how OLS is useful for the promotion of science teachers' innovative work behavior.

At first, we investigated the latent construct *organization leadership support* on the basis of the TIMSS 2015 questionnaire. An EFA was carried out for investigating the variables of organization leadership support. The EFA revealed the group of questions related to organization leadership support: teachers' understanding of the school's curricular goals, collaboration between school leadership and teachers to plan instruction, amount of instructional support provided to teachers by school leadership, and school leadership's support for teachers' professional development. A CFA was performed in order to confirm EFA results about organization leadership support. The five observed variables are responses to the four OLS questions. The strength of the CFA path analysis lies in its ability to reveal the relationships between the observed variables and the latent variable. We observed the highest loading for the variable *collaboration between school leadership and teachers to plan instruction*. This variable expresses relational leadership. Scholars state that "leadership is not based only on the traits of the leader but it is the social process that occurs between followers and the leader. i.e. leadership is a relational process" (Akram et al., 2016). Our research confirms the role of relational leadership on the latent variable OLS ($\beta = 0.761$, $p < 0.001$).

We analyzed the influence of organization leadership support (OLS) on science teachers' innovative work behavior. The role of leadership in innovation is explored not only in education. Akram et al. (2016) investigated the effect of relational leadership on employees' innovative work behavior in the information technology industry of China. Scholars revealed that relational leadership helps employees to cultivate and demonstrate innovative work behavior at each stage of the idea generation process. Results of our research in the field of education confirm our main conclusion about the influence of leadership on the innovation process in technology industries in different cultures: "In order to improve the employee innovative work behavior, organizations need to promote relational leadership among their leaders" (Akram et al., 2016).

The moderator role of OLS for innovative work behaviors in non-educational organizations was analyzed by Mokhber et al. (2018). They found a positive relationship between transformational leadership and organizational innovation in their study of the main functions or traits associated with innovative work behavior: idea generating, risk taking, and decision-making. Scholars state "that transformational leaders might not only promote innovative activity within the organization but also ensure the market success of the innovations" (Mokhber et al., 2018, p. 108). We have not addressed the role of transformational leaders in the innovation process within an organization. We did not consider that type of leadership in our study. However, our study confirmed the following trends: OLS has an influence on the innovative work behavior of science teachers. The results of our research were broader in scope and cannot be confined to just one type of leadership. We believe further research is needed to study in depth the different types of leadership: transversal, shared, and so on.

### 3.5.5. Conclusions

The findings drawn from our empirical analysis of the role of OLS in science teachers' innovative work behavior abilities showed that OLS influenced in different ways the science teachers' innovative work behavior.

It was found that OLS had more influence on science teachers' ability to generate and share new ideas and less influence on their ability to apply new ideas in practice. OLS had a statistically significant influence on science teachers' abilities to work together in trying out new ideas and sharing what they have learned.

## 3.6. The associations between professional development, professional, demographic factors and the innovative work behavior of science teachers

> *This section analyzes the associations between professional development, professional and personal demographic factors, and the innovative work behavior of science teachers based on data collected from three countries: Singapore, Lithuania, and Sweden.*

### 3.6.1. Introduction

The quality of teacher's professional development depends on initiatives such as bringing about and implementing changes successfully in the field of education and improved teaching practices that, in turn, contribute to higher achievements of students (Darling-Hammond et al., 2017; Hargreaves, 2006; Hargreaves, 2019; Karlberg & Bezzina, 2020). Researchers also note that specific abilities of teachers and school practices, including collaborative work and a supportive environment (Campbell, 2019; Klaeijsen, Vermeulen & Martens, 2018), increase chances for the school system to pilot innovative practices and strive for higher-quality education.

Innovative practice in science education depends on science teachers' innovative work behavior. Researchers regard innovative work behavior as the ability to create and implement something new in the existing work system (Aziah & Al Amin, 2018; Chang, 2018; Pudjiarti & Hutomo, 2020). Innovative work behavior is reflected by a series of activities in which individuals generate novel ideas, solve practical problems at work, and share new ideas (Rogers, 2003; Sun & Huang, 2019).

Innovative practice depends on teachers' professionalism. In general, teachers gain professional experience in a process that begins with pre-service training and continues with in-service training (Kaya & Gödek, 2016). Brekelmans, Wubbels, and van Tartwijk (2005) discussing about role of teachers' experience in education revealed two factors: time for acquiring the skills needed to tackle challenges and enthusiasm toward bringing in changes. Less experienced teachers and those in the early stage of their careers may take a few years before they acquire the skills to manage challenges and create innovations; however, only such teachers demonstrate more enthusiasm and willingness to face and overcome challenges.

Hargreaves (2005) argues that experienced teachers, on the other hand, feel more relaxed and comfortable with a stable, predictable, and routine type of professional activity. In fact, more experienced teachers become resistant toward change at the end of their career (Hargreaves, 2005). Given this, it is more likely for teachers late in their career to lose their energy and show a decline in their interest to making improvements in their work practices than their younger colleagues. Hargreaves' (2005) findings suggest that less experienced teachers are more enthusiastic about making and implementing change initiatives in their own classrooms—a place they believe to be the most ideal where they can best make a difference (Hargreaves, 2005).

Research on teachers' professional development is developed in multiple ways responding to challenges posed by the complexities of society and education. The experience gained from pre-service teacher education is not enough for an effective teaching-learning process; as the teacher is a continuous learner, they need to engage in in-service teacher education programs (Bada & Prasadh, 2019). Specifically, much attention has been paid to effective professional development of teachers.

Effective professional development of teachers leads to changes in teaching practices and student learning outcomes (Chaudhuri, McCormick, & Lewis, 2019; EL-Deghaidy, Mansour, Aldahmash, & Alshamrani, 2015; Lownsbrough, 2020; Sims, & Fletcher-Wood, 2021; Smith, Ralston, Naegele, & Waggoner, 2020). Researchers summarizing their reviews of as many as 35 methodologically rigorous studies pointed out 7 features of effective professional development: It is content focused, incorporates active learning, supports collaboration, uses models of effective practice, provides coaching and expert support, offers feedback and reflection, and is of sustained duration (Darling-Hammond et al., 2017). In this study, we analyze content-focused development and sustained duration.

The aim of this study is to reveal the role of professional development content, duration, teachers' gender, and teaching experience in their innovative work behavior.

In seeking to fulfil our study objectives, we performed an in-depth analysis of teacher training and professional development in Singapore, Lithuania, and Sweden.

*Teacher training in Lithuania*

Qualifications that allow one to teach at a school may be acquired at higher education institutions. A minimum requirement is bachelor's degree (first-cycle studies). Universities provide bachelor's studies (240 ECTS in 4 years), while colleges offer professional bachelor's studies (210 ECTS in 3 years; Eurydice, 2019). A pedagogical internship is required for newly graduated teachers to go through before starting work at the school (Kelly, Centurino, Martin, & Mullis, 2020). Teachers are encouraged to obtain master's degrees, which make them eligible to teach one more subject or allow taking on a more pedagogical role, (Shewbridge et al., 2016).

Those with a higher education degree may also enter teaching profession through the program called Teach for All (*Renkuosi mokyti!* in Lithuania), which is an alternative pathway to the teaching profession (Eurydice, 2019).

Researchers showed that the subject and curricular matter received more emphasis before pedagogy in the initial phase of teacher education (Shewbridge et al., 2016). However, according to TALIS 2018 results, 82% of teachers in Lithuania received education in both subject area and specialized pedagogical knowledge, which is more than the OECD average ; OECD, 2019).

However, making available a balanced mix of content in the initial phase of teacher education still remains a challenge. Other challenges include making connections between initial teacher education and professional development, as well as raising attractiveness of the teaching profession in the society (Shewbridge et al., 2016).

*Teacher training in Sweden*

In Sweden, universities and university colleges provide teacher education. The duration of study varies between 3 and 5.5 years. Since 2011, bachelor and master degrees have been divided into four types of professional degrees:

- Preschool education—210 ECTS;
- Primary school education—240 ECTS;
- Subject education—270 ECTS, out of which 195 are designated to the subject area;
- Vocational education—90 ECTS and relevant vocational knowledge required for admission (Eurydice, 2020).

The period of studies also depends on the school level the teacher-to-be intends to teach, as well as the subject. For example, to gain a degree in subject education for upper secondary schools, one should take 300 or 330 ECTS. The studies toward a degree in primary education for working in out-of-school care requires 180 ECTS, which is the minimum length of teacher education. There is no official requirement for a qualification for one to work at preschools (Eurydice, 2020).

In Sweden, around 23% of teacher-students have had the experience of studying abroad (OECD, 2019).

The content of the study program includes subject knowledge and didactics, educational science, as well as internship placement (Mullis et al., 2016).

As for alternative pathways to teacher qualification, those holding a higher education degree can receive a teacher qualification for 1.5 years of supplementary teacher training (Eurydice, 2020). Also, there is a mechanism in place for recognizing foreign teacher qualifications (Skolverket, 2021). The Swedish policy in the area is currently aimed at widening the pool of potential teachers by supporting different ways of entry into the profession (Schleicher, 2019). Government agencies also work for increasing the social image of the teacher profession.

*Teacher training in Singapore*

The Ministry of Education Candidates recruits directly from a pool of high-achieving graduates for filling public school teaching positions. Successful candidates enter the initial teaching education provided by the National Institute of Education (NIE)—the only institution for teacher education. NIE offers bachelors and graduate programs (NIE, 2021). According to TIMSS 2015 Encyclopaedia, candidates hold university degrees in various subjects, along with a small proportion of those who are high school (A-level) graduates (Mullis et al., 2016). The admission requirement, stated at the NIE website, is A-level high school diploma with high scores in subjects that candidates intend to study at the institute (NIE, 2021).

Teacher education programs offered at NIE are developed in line with the content of the national curriculum for school education. A teacher internship with a duration of minimum 10 weeks is required (Mullis et al., 2016). According to TALIS 2018 results, more than 95% of teachers have completed an internship (OECD, 2019).

Teacher education in Singapore is a highly competitive field due to the benefits that teaching profession offers (Mullis et al., 2016). The government constantly adjusts the salaries of teachers to match the scale of salaries offered in the private sector, aiming to keep the profession prestigious (Schleicher, 2019).

The system of education as well as teacher education in Singapore is often praised in the literature due to the country's high rates in international surveys, including TALIS and PISA.

## Continuous teacher education in Lithuania

Teachers are obliged to take part in continuous professional development (CPD), with five days per year reserved for it according to the legislation (Eurydice, 2019; OECD, 2019). CPD is necessary for retaining employment and promotion (OECD, 2019; Shewbridge et al., 2016). Lithuania has one of the highest shares of teacher participation in CPD activities across the OECD countries (OECD, 2019). Participation in PCD is a prerequisite for receiving a qualification category. Currently, the ICT skills and skills for teaching for special educational needs are emphasized in the training activities for teachers. Teachers may also demand a performance evaluation (Shewbridge et al., 2016).

Professional development is provided for a fee by governmental and nongovernmental institutions. The fee may be paid by teachers, or schools can cover it from the student basket budget. The OECD review form 2016 (Shewbridge et al., 2016) identified such challenges in teacher's professional development:

- it is "rather fragmented" with uneven budget allocation between municipalities; the training providers also operate in free market conditions. (p. 136)
- it is undertaken on individual basis according to teachers' perspective, without considering school strategic needs for competence development and a targeted benefit for the school learners.

## Continuous teacher education—Sweden

Professional development for teachers in Sweden is not obligatory and is not linked to career promotion (Eurydice, 2020). Currently, 104 hours of teachers' working time per school year is reserved for professional development activities. This time is negotiated by teachers' unions with local authorities (Eurydice, 2020).

The central government does not issue directives about teacher development activities, but local authorities (communes) are vested with the formal responsibility of providing teachers with opportunities for professional development. Local authorities receive financial grants from the Swedish National Agency for Education for development of teachers' skills in national priority areas (Eurydice, 2020).

Publicly funded CPD activities are available from higher education institutions, regional development centers, trade unions, state authorities, independent educational companies, and nongovernmental organizations.

An important part of professional development of Swedish teachers is the focus on the capacity of teachers to teach in diverse, multicultural, and multilingual environments, as well as teaching Swedish as a second language (OECD, 2019).

One of the challenges is that collaborative learning is not widely practiced in professional development in Sweden, leaving teachers to follow courses individually, which is not a preferred option for all the teachers (Karlberg & Bezzina, 2020).

### Continuous teacher education—Singapore

All practicing teachers undergo over 100 hours of professional development per year. CPD is not obligatory by law. The schools in Singapore have a common vision on professional development, so that the CPD activities are incorporated into teachers' daily practice (OECD, 2019).

Professional development courses are provided by the NIE and the Ministry of Education. The courses are offered in specialized subject and pedagogical areas. Experiential learning programs for teachers in cooperation with private sector are available since 2003. Besides, around 65% of teachers in Singapore participate in networks of teachers for professional development (OECD, 2019). Teachers are also offered opportunities to develop and demonstrate their leadership skills from the beginning of their careers, partly to prepare them for school leadership roles in the future. A specially designed 6-month preservice program, Leaders in Education (LEP), for new principals was introduced in 2001 (OECD, 2019).

Singapore has a system that connects teacher professional development and career progress in three directions: teaching, leadership, and expertise for education development. The system also includes career path planning, remuneration plan, and performance evaluation (Schleicher, 2019).

CPD plays a significant role in the teaching career, since those for whom training activities had an impact on their teaching reported being more satisfied with their jobs than those who did not report the training to be effective (OECD, 2019).

The challenges, according to Bautista, Wong, and Gopinathan (2015), include high workload of teachers; participation in professional development (PD) not allowed for all but allowed for only one subject that teachers teach; and the prevailing culture that focuses on high test scores of students (testing culture), so that teachers continue to teach students for tests, instead of applying the various methods they learned from PD training.

The main differences and similarities between the countries are presented in Tab. 3.6.1.

**Tab. 3.6.1:** Main features of initial teacher education in the three countries chosen for the study: Lithuania, Sweden, and Singapore

| | Lithuania | Sweden | Singapore |
|---|---|---|---|
| Education can be acquired at | HEI: Universities and colleges | HEI: universities and university colleges | National Institute of Education only |
| Minimum degree in teaching | Bachelor's/ professional bachelor's (first cycle) | Professional degree in education (first cycle) | Bachelor's (first cycle) |
| Entry requirements | Final high school grades; Motivation | Final high school grades or corresponding knowledge | Final high school grades |
| Minimum period of study | 3 years | 3 years | 3.5 years |
| Studies involve both the subject and pedagogy aspects | yes | yes | Yes |
| Internship/practicum is part of studies | yes | yes | Yes |
| Alternative pathway available | yes | yes, different options | /no info available |
| Social prestige of teaching profession | Not high | Not high | High |

The level of the highest education completed by the teachers varies across countries. In Sweden, the majority hold a qualification equivalent to an ISCED level 7 degree, while in Lithuania the majority of lower-secondary teachers hold a bachelor's degree (level 6). In Singapore, the bachelor's degree (level 6) prevails substantially (OECD, 2019).

The difference can be explained by the fact that Sweden has launched special professional degrees for teachers since 2011, which is a system used for regulated fields of study, such as law and medicine. Thus, teachers for secondary and upper secondary schools have to complete over 4 years of study, which corresponds to the second cycle—ISCED level 7 degree.

At the same time, a consecutive master's degree in Lithuania provides additional qualification for teachers.

**Tab. 3.6.2:** TALIS 2018 indicators for initial teacher education and novice period in the countries chosen for the study: Lithuania, Sweden, and Singapore

| | Lithuania | Sweden | Singapore | OECD average |
|---|---|---|---|---|
| During their initial education and training, teachers were instructed on subject content, pedagogy, and classroom practice | 82% | 85% | 89% | 79% |
| Teachers participated in a kind of formal or informal induction when they joined their current school | 21% | 30% | 85% | 42% |
| Novice teachers had a mentor assigned | 9% | 17% | 54% | 22% |

*Based on country notes: Lithuania, Sweden, Singapore in OECD (2019), TALIS 2018 Results (Volume I): Teachers and School Leaders as Lifelong Learners, TALIS, OECD Publishing, Paris.*

The recruitment and education of teachers are a centralized and regulated process in Singapore. Sweden and Lithuania have flexible options for candidates receiving teacher education.

In Sweden, initial teacher education is highly flexible: Students may choose how long they study at the university according to the school level they would like to teach in the future. It takes less time to acquire a primary school teacher qualification than a secondary school teacher qualification.

Swedish authorities are working at broadening the pool of candidates interested in teaching by offering different ways of entry into the profession, since there is a demand for teaching staff.

Lithuania has a smaller number of novice teachers, as demonstrated by TALIS 2018—7% in Lithuania compared to 19% across the OECD countries.

The need for connecting initial teacher education with professional development in Lithuania is supported by the fact that only one-fifth of the newly graduated teachers receive induction activities to transfer into working life at schools.

As for gender distribution in teaching profession, TALIS 2018 reported 85% of female teachers in Lithuania, 66% in Sweden, and 64% in Singapore (OECD average—68%; OECD, 2019).

The main differences and similarities between the countries are presented in Tab. 3.6.3.

**Tab. 3.6.3:** Main features of continuous teacher education in the countries chosen for the study: Lithuania, Sweden, and Singapore

|  | Lithuania | Sweden | Singapore |
|---|---|---|---|
| CPD is compulsory by law | Yes | No | No |
| Legally designated time for CPD | 5 days/year (~ 40 hours) | 104 hours/year | 100 hours/year |
| CPD facilitates career progress for teachers | Yes | Not necessarily | Yes |
| Nature of CPD providers | Public and private | Public and private | Public |
| Access to publicly funded CPD available | yes | yes | Yes |

CPD in Lithuania is mandatory but not centralized. In Singapore, however, it is centralized and embedded in teaching practice at the school as well as in career development. In Sweden, CPD is rather flexible for teachers and has no direct link to career progress. At the same time, teachers in Sweden are less likely to state that training activities had a positive impact on their teaching than are those in the other two countries (Tab. 3.6.4.)

**Tab. 3.6.4:** TALIS 2018 indicators for continuous teacher education in the countries chosen for the study: Lithuania, Sweden, and Singapore

|  | Lithuania | Sweden | Singapore | OECD average |
|---|---|---|---|---|
| Teachers participated in courses and seminars in in-service training | 97% | 74% | 94% | — |
| Teachers participated in training based on peer learning and coaching | 69% | 47% | 77% | — |
| An elaboration of professional development or a training plan offered after teacher assessment for percentage of teachers | 97% | 96% | 100% | 90% |
| A mentor is appointed after teacher assessment for percentage of teachers | 64% | 89% | 100% | 71% |
| Teachers for whom training had a positive impact on practice | 89% | 73% | 91% | 82% |
| Percentage of teachers who took part in in-service training activities in the year before the survey | 99% | 95% | 98% | 94% |

Based on country notes: Lithuania, Sweden, and Singapore in OECD (2019), *TALIS 2018 Results (Volume I): Teachers and School Leaders as Lifelong Learners*, TALIS, OECD Publishing, Paris.

Teachers in Sweden and Singapore are entitled to more time for in-service training than teachers in Lithuania.

Mentorship practices are the most widespread in Singapore; especially novice teachers are supervised by mentors (Mullis et al., 2016). They also receive less direct teaching instruction hours (OECD, 2019).

Peer learning in Lithuania and Sweden is less emphasized than in Singapore.

In Lithuania and Sweden, most teachers require additional training in ICT skills, while teachers in Singapore lack skills for working with pupils with special needs (OECD, 2019).

### 3.6.2. Method of research

We addressed the core research theme of our study using the database of International Mathematics and Science Study (TIMSS, 2015). We selected the TIMSS 2015 data because the TIMSS 2015 questionnaire included questions related to innovative work activities of science teachers.

To identify factors influencing innovative work activities of science teachers, we used the TIMSS 2015 database on three different countries: Lithuania, Sweden, and Singapore. These countries were selected based on student achievement in TIMSS 2015. The research samples (countries) were selected on the principle of probability systematic sampling using distribution of science achievement (Mullis et al., 2016). Singapore is the country leading in terms of student achievement in science. Lithuanian and Sweden students' science achievements are close to average on achievement distribution (Mullis et al., 2016).

The TIMSS questionnaire allowed us to investigate the curriculum content of professional development training given to science teachers (Tab. 3.6.5). Singapore teachers are most often involved in professional development in areas such as science pedagogy, science content, and science curriculum. Science teachers from Lithuania are most often involved in professional development in science content and science assessment. Teachers from Sweden go through professional development in science curriculum and science content. The percentage of teachers' participation in professional development with different content is very similar between Sweden and Lithuania. The difference in the rate of participation is about 12% (Table 3.6.5). In Singapore, the percentage of participation differed substantially between the most popular (science pedagogy) and the least popular training topics (addressing individual student's need).

**Tab. 3.6.5:** The percentage of science teachers' participation in professional development: the aspect of content

| TIMSSS 2015 questions about professional development | Singapore | Lithuania | Sweden |
|---|---|---|---|
| Science content | 69.8 | 63.8 | 32.4 |
| Science pedagogy | **90.4** | 55.7 | 27.1 |
| Science curriculum | 67.1 | 55.7 | **34.3** |
| Integrating information and communication technology | 66.8 | 62.2 | 26.8 |
| Improving students critical thinking or inquiry skills | 65.6 | **44.7** | **22.2** |
| Science assessment | 59.6 | 60.0 | 31.4 |
| Addressing individual students' need | **39.8** | 51.7 | 30.1 |

The mean and median of the numbers of years the science teachers have been teaching vary from country to country (Tab. 3.6.6). According to the TIMSS 2015 data, the highest median of number of teaching years is in Lithuania and the lowest in Singapore. A higher percentage of male teachers participated in the TIMSS 2015 study in Singapore (38.3%) and Sweden (40.7%). And only 12% of males participated in the TIMSS 2015 study in Lithuania (Table 3.6.6).

**Tab. 3.6.6:** Descriptive statistic of science teachers: sample, the number of years the teacher has been teaching, and the gender of the teacher

|  | Singapore | Lithuania | Sweden |
|---|---|---|---|
| Sample | 334 | 941 | 647 |
| Number of years the teacher has been teaching (mean) | 8.64 | 24.54 | 12.92 |
| Number of years the teacher has been teaching (median) | **6.00** | **25.00** | 12.00 |
| Gender: male | **38.3** | 12.7 | **40.7** |
| Gender: female | 61.7 | 87.3 | 59.3 |

The TIMSS 2015 questionnaire allowed us to measure the duration (number of hours) of science teachers' professional development (Tab. 3.6.7). We observed that the duration of teachers' participation in professional development varied vastly between Singapore and in Sweden (Tab. 3.6.7). Teachers from Sweden are more involved in short-term professional development, whereas teachers from Singapore and Lithuania undergo professional development for longer durations (Tab. 3.6.7).

**Tab. 3.6.7:** The time science teachers spent in professional development in the last two years: Singapore, Lithuania, and Sweden

|  | Singapore | Lithuania | Sweden |
|---|---|---|---|
| None | 2.1 | 2.9 | 33.5 |
| Less than 6 hours | 11.4 | 7.3 | 27.0 |
| 6–15 hours | 24.6 | 31.5 | 22.5 |
| 16–35 hours | 24.9 | 30.3 | 7.2 |
| More than 35 hours | 36.8 | 28.0 | 9.7 |

*Mathematical models*

We created four mathematical models to identify the influence of professional development and professional and demographic factors on the innovative work behavior of science teachers from Singapore, Lithuania, and Sweden.

$TONI = f(DPD, CPD_i, GEN, PE)$ (1).
$CHAMP = f(DPD, CPD_i, GEN, PE)$ (2).
$PPS = f(DPD, CPD_i, GEN, PE)$ (3).
$SHARE = f(DPD, CPD_i, GEN, PE)$ (4).

Acronyms and abbreviations used for the names of the variables in the models are as follows:

TONI—Try out new ideas
CHAMP—Championing new ideas
PPS—Promotion of the problem-solving
SHARE—Sharing new ideas
DPD—Duration of professional development
$CPD_i$—Content of professional development
GEN—Gender of the teacher
PE—Professional experience (years the teacher has been teaching).

$CPD_i$ is a complex variable and includes different training topics (Tab. 3.6.8). CPD variable's ranks are 1 = Yes; 2 = No.

Tab. 3.6.8: Contents of science teachers' professional development

| Question code (CPD$_i$) | The content of professional development |
| --- | --- |
| BTBS 23A | Science content |
| BTBS 23B | Science pedagogy |
| BTBS 23C | Science curriculum |
| BTBS 23D | Integrating information technology into science |
| BTBS 23E | Improving students' critical thinking or inquiry skills |
| BTBS 23F | Science assessment |
| BTBS 23G | Addressing individual student's needs |

The OLR process involves checking four assumptions. The first assumption states that dependent variable should be measured at the ordinal level. Three dependent variables (TONI, CHAMP, SHARE) had four ordered categories: *Very often, Often, Sometimes, Never,* or *Almost never*. Applying percentage frequencies, we noticed that the dependent variables (TONI, CHAMP, SHARE) acquired only very small values on the response categories *never* or *almost never*, so it is more appropriate to use a categorical variable with fewer categories. We created a new dependent variable with three categories: 1 = *very often* and often, 2 = *sometimes*, 3 = *never*; we created two (k-1) equations, each with a different intercept but with the same *b* coefficients (slopes) for predictor variables. It means the effects of independent variables are the same for each level of the dependent variable. We tested this condition with the "test of parallel lines assumption."

The dependent variable *promotion of problem-solving* had four categories in TIMSS 2015: *every or almost every lesson, about half the lessons, some lessons,* and *never*. Applying percentage frequencies, we noticed that dependent variables (PPS) acquired only very small values in the response category *Never*. That is why we created a new dependent variable *PPS* with three categories: *every or almost every lesson, about half the lessons, some lessons,* and *never*.

The second assumption states that one or more independent variables must be continuous, ordinal, or categorical (including dichotomous variables). However, ordinal independent variables must be treated as being either continuous or categorical. We separated independent variables (DPD, CPD$_i$, PE) into covariates and factors. The continuous variable—professional experience (PE)—has been assigned to covariates. We treated the independent ordinal variable CPD$_i$—content of professional development—as categorical (dichotomous) and assigned it to covariates. We created a new nominal variable DPD with two categories: 1 = short professional development (up to 15 hours) and 2 = long professional development (more than 16 hours). The categorical variable DPD has been assigned to factors.

### 3.6.3. Results of research

*The role of professional development, professional and demographic factors in trying out new ideas in science education*

Based on the first model (TONI) we tested hypothesis $H_1$: The science teacher's willingness to try out new ideas collaboratively is influenced by the duration of professional development, gender of teacher, years of teaching, content of professional development, and duration of professional development.

Before we start looking at the effects of explanatory variable in the first model, it is important to determine whether the model improves the possibility of predicting outcomes. We compare a model without any explanatory variables (the baseline or "intercept only" model) with the model with all the explanatory variables (the "final" model—this would normally have several explanatory variables but at the moment it just contains level of education; Table 3.6.9). We ensured model's fit with the data by creating the study parameters based on the TIMMS 2015 database for the three countries—Lithuania, Sweden, and Singapore (Tab. 3.6.9). The significant $\chi^2$ statistic ($p < 0.05$) indicates that the final model gives a significant improvement over the baseline intercept-only model. The small $p$ value from the model-fitting test, <.05, makes it possible to conclude that at least one of the regression coefficients in the model is not equal to zero.

**Tab. 3.6.9:** Model (1) fitting information from an ordinal regression analysis: to explain science teachers trying out new ideas in science education

| Country | Dependent variable | Model | −2 Log Likelihood | $\chi^2$ | df | Sig. | Pearson's $\chi^2$ statistic |
|---|---|---|---|---|---|---|---|
| Lithuania | TONI | Intercept only | 1473.368 | | | | |
| | | Final | 1428.983 | 44.385 | 10 | .000 | .067 |
| Sweden | TONI | Intercept only | 1937.894 | | | | |
| | | Final | 1901.419 | 53.979 | 10 | .000 | .193 |
| Singapore | TONI | Intercept only | 1297.739 | | | | |
| | | Final | 1287.095 | 10.644 | 10 | .040 | .692 |

The important parameter in the output is Pearson's $\chi^2$ statistic for the model. These statistics are intended to test whether the observed data are consistent with the fitted model. If we do not reject this hypothesis ($p > 0.05$), then we can conclude that the data and the model predictions are similar and that we have a good model (Tab. 3.6.9).

For the predictor CPD, Wald coefficient has two statistical values: one for science curriculum training (BTBS 23C; $p = 0.006 < 0.05$) and one for improving students' critical thinking or inquiry skills (BTBS 23E; $p = 0.04 < 0.05$; Tab. 3.6.10). For those Lithuanian science teachers who had science curriculum as PD content, the odds to try out new ideas were 1.487 (95% CI: 1.107–1.998) times that of those who did not have this PD content: $\chi^2(1) = 0.418$, $p = 0.006$. For those Lithuanian science teachers who had improving students' critical thinking or inquiry skills as PD content, the odds to try out new ideas were 1.557 (95% CI: 1.170–2.073) times that of those who did not have this PD content: $\chi^2(1) = 0.418$, $p = 0.004$.

The mathematical model on the data from Lithuania for the logarithms (logit functions) of the dependent variable is created based on the first model, TONI:

$$\ln\frac{P(TONI \le i)}{P(TONI > i)} = \begin{cases} 1.786, \text{ when } i = 1 \\ 5.525, \text{ when, } i = 2 \end{cases} + 0.572(\text{GEN}) + 0.418(\text{BTBS23C}) + 0.418(\text{BTBS23E})$$

(1a)

The positive signs indicate that the predictors BTBS 23C and BTBS 23E have positive effects on science teachers' activity of trying out new ideas.

**Tab. 3.6.10:** Parameter estimates to explain science teachers' activity of trying out new ideas in science education: based on the data for Lithuania

|  |  | Estimate | Std. Error | Wald | df | Sig. | 95% Confidence Interval | |
|---|---|---|---|---|---|---|---|---|
|  |  |  |  |  |  |  | Lower Bound | Upper Bound |
| Threshold | [TONI = 1.00] | 1.786 | .496 | 12.979 | 1 | .000 | .814 | 2.757 |
|  | [TONI = 2.00] | 5.525 | .539 | 105.154 | 1 | .000 | 4.469 | 6.582 |
| Location | GEN | -.004 | .006 | .315 | 1 | .575 | -.016 | .009 |
|  | PE | -.572 | .217 | 6.981 | 1 | .008 | -.148 | .997 |
|  | BTBS 23A | .218 | .157 | 1.932 | 1 | .164 | -.089 | .525 |
|  | BTBS 23B | -.055 | .146 | .142 | 1 | .706 | -.340 | .230 |
|  | BTBS 23C | .418 | .152 | 7.598 | 1 | .006 | .121 | .715 |
|  | BTBS 23D | -.068 | .144 | .224 | 1 | .636 | -.350 | .214 |
|  | BTBS 23E | .418 | .146 | 8.194 | 1 | .004 | .132 | .704 |
|  | BTBS 23F | .207 | .150 | 1.890 | 1 | .169 | -.088 | .501 |
|  | BTBS 23G | .251 | .142 | 3.113 | 1 | .078 | -.028 | .531 |
|  | [DPD = 1.00] | -.034 | .152 | .050 | 1 | .823 | -.332 | .264 |
|  | [DPD = 2.00] | 0ª | . | . | 0 | . | . | . |

Link function: Logit.
a. This parameter is set to zero because it is redundant.

Wald statistics on the data from Lithuania revealed small but statistically significant negative influence of the number of years the teacher has been teaching (PE) on science teachers' activity of trying out new ideas (Tab. 3.6.10). For those Lithuanian science teachers who had more experience, the odds of trying out new ideas were 0.997 (95% CI: 0.985–1.010) times that of those who had less experience with this PD content: $\chi^2(1) = -0.572$, $p = 0.008$. An odds ratio < 1 suggests a decreasing probability of dependent variable with increasing values on an independent variable. This result is due to the data coding order in the TIMSS study. The higher ranks of TONI questionnaire were numbered in lower numbers.

We have performed an OLR analysis based on Wald statistics on the data from Sweden (Tab. 3.6.11). Wald coefficients indicate that there is no statistically significant association between science teachers' trying out new ideas and their gender. The obtained result differs from the result obtained on the basis of the Lithuanian database. The distribution of teachers by gender in Lithuania and Sweden is similar nevertheless.

We get a statistically significant confirmation that the probability of science teachers' trying out new ideas in science education depended in a statistically significant manner on the content of professional development (Tab. 3.6.11). For the predictor CPD, the Wald coefficient has two statistically significant values: one for science curriculum training (BTBS 23C; $p$ = 0.043 < 0.05) and one for integrating information technology into science (BTBS 23D; $p$ = 0.006 < 0.05; Tab. 3.6.11). For those Swedish science teachers who had science curriculum as PD content, the odds of championing new ideas were 1.004 (95% CI: 0.682–1.596) times that of those who did not have this PD content, $\chi^2(1)$ = 0.387, $p$ = 0.043. For those Swedish science teachers who had the integrating information technology into science as PD content, the odds of championing new ideas were 1.691 (95% CI: 1.151–2.248) times that of those who did not have this PD content, $\chi^2(1)$ = 0.562, $p$ = 0.006.

Tab. 3.6.11: Parameter estimates to explain science teachers' activity of trying out new ideas in science education: based on the data for Sweden

| | | Estimate | Std. Error | Wald | df | Sig. | 95% Confidence Interval | |
|---|---|---|---|---|---|---|---|---|
| | | | | | | | Lower Bound | Upper Bound |
| Threshold | [TONI = 1.00] | 2.460 | .555 | 19.642 | 1 | .000 | 1.372 | 3.549 |
| | [TONI = 2.00] | 4.901 | .581 | 71.183 | 1 | .000 | 3.763 | 6.040 |
| Location | GEN | -.005 | .009 | .348 | 1 | .555 | -.023 | .012 |
| | PE | .212 | .165 | 1.646 | 1 | .199 | -.112 | .535 |
| | BTBS 23A | .276 | .212 | 1.692 | 1 | .193 | -.140 | .693 |
| | BTBS 23B | .370 | .240 | 2.382 | 1 | .123 | -.100 | .841 |
| | BTBS 23C | .387 | .208 | 3.459 | 1 | .043 | -.794 | .021 |
| | BTBS 23D | .562 | .206 | 7.486 | 1 | .006 | .160 | .965 |
| | BTBS 23E | .246 | .221 | 1.247 | 1 | .264 | -.186 | .678 |
| | BTBS 23F | .273 | .195 | 1.954 | 1 | .162 | -.110 | .656 |
| | BTBS 23G | -.041 | .191 | .047 | 1 | .829 | -.416 | .334 |
| | [DPD = 1.00] | .284 | .273 | 1.082 | 1 | .298 | -.251 | .818 |
| | [DPD = 2.00] | 0ª | . | . | 0 | . | . | . |
| Link function: Logit. | | | | | | | | |

a. This parameter is set to zero because it is redundant.

The threshold coefficients represent the intercepts, specifically the point (in terms of a logit) where science teachers' activity of trying out new ideas might fall under two categories (Tab. 3.6.11). The mathematical model for logit functions of the dependent variable is a part of the TONI model which is created based on the data for Sweden:

$$\ln\frac{P(TONI \leq i)}{P(TONI > i)} = \begin{cases} 2.460, when\ i = 1 \\ 4.901, when,\ i = 2 \end{cases} + 0.387(BTBS23C) + 0.562(BTBS23D) \quad (1b)$$

All coefficients in the mathematical model based on data for Sweden (1b) have positive signs. The positive sign indicates that two predictors have positive effects on science teachers' activity of generating new ideas in education: professional development of science content curriculum, and integrating information technology into science. The results of parallel lines test confirm the null hypothesis that the TONI model has one set of coefficients ($p > 0.05$; Tab. 3.6.13).

We analyzed the data for Singapore to ascertain the influence of different predictors (DPD, CPD, GEN, PE) on science teachers' activity of generating new ideas based on the first model (1). The results of Wald statistics did not reveal a statistically significant influence of the predictors DPD, CPD, GEN, and PE on the dependent variable GNI (Tab. 3.6.12).

Tab. 3.6.12: Parameter estimates for the TONI model: Singapore

|  |  | Estimate | Std. Error | Wald | Df | Sig. | 95% Confidence Interval | |
|---|---|---|---|---|---|---|---|---|
|  |  |  |  |  |  |  | Lower Bound | Upper Bound |
| Threshold | [TONI = 1.00] | 1.234 | .693 | 3.168 | 1 | .025 | -.125 | 2.594 |
|  | [TONI = 2.00] | 4.255 | .744 | 32.691 | 1 | .000 | 2.796 | 5.713 |
| Location | PE | -.311 | .291 | 1.149 | 1 | .284 | -.881 | .258 |
|  | BTBS 23A | .127 | .267 | .228 | 1 | .633 | -.396 | .651 |
|  | BTBS 23B | .089 | .411 | .047 | 1 | .829 | -.718 | .895 |
|  | BTBS 23C | .270 | .277 | .949 | 1 | .330 | -.273 | .812 |
|  | BTBS 23D | -.007 | .245 | .001 | 1 | .979 | -.486 | .473 |
|  | BTBS 23E | .270 | .260 | 1.080 | 1 | .299 | -.239 | .780 |
|  | BTBS 23F | .213 | .245 | .752 | 1 | .386 | -.268 | .693 |
|  | BTBS 23G | .202 | .241 | .705 | 1 | .401 | -.270 | .675 |
|  | GEN | .016 | .017 | .812 | 1 | .368 | -.018 | .050 |
|  | [DPD = 1.00] | .354 | .173 | 1.001 | 1 | .222 | -.351 | .418 |
|  | [DPD = 2.00] | 0a | . | . | 0 | . | . | . |

Link function: Logit.
a. This parameter is set to zero because it is redundant.

The "test of parallel lines" compares the ordinal model that has one set of coefficients for all thresholds (null hypothesis) to a model with a separate set of coefficients for each threshold (general). We are led to reject the assumption of proportional odds if the general model gives a significantly better fit to the data than the ordinal (proportional odds) model ($p < 0.05$). The results of parallel lines test confirm the null hypothesis that the TONI model has one set of coefficients based on the data for Lithuania, Sweden, and Singapore (Tab. 3.6.13).

**Tab. 3.6.13:** Test of parallel lines: to explain science teachers' activity of trying out new ideas in science education: based on the data from Lithuania, Sweden, and Singapore

| Country | Model | | -2 Log Likelihood | $\chi^2$ | df | Sig. |
|---|---|---|---|---|---|---|
| Lithuania | TONI | Null hypothesis | 1391.469 | | | |
| | | General | 1384.963 | 6.506 | 10 | .771 |
| Sweden | TONI | Null hypothesis | 1010.796 | | | |
| | | General | 982.747 | 8.049 | 10 | .512 |
| Singapore | TONI | Null hypothesis | 287.095 | | | |
| | | General | 285.745 | 1.350 | 7 | .987 |

We compared the results of our OLR analysis based on data for the three countries—Lithuania, Sweden, and Singapore. The analysis of Singapore data showed that predictors DPD, $CPD_i$, GEN, and PE did not statistically significantly affect science teachers' activity of trying out new ideas. In only one case (based on Lithuanian data) there was a statistically significant influence of professional experience (PE) on the activity of generating new ideas. This trend has not been established based on data for countries other than Lithuania, Singapore, and Sweden.

Summarizing the results of our OLR analysis of the data for Lithuania, Sweden, and Singapore on the influence of CPD predictor on science teachers' activity of generating new ideas, we noticed that the CPD component "science curriculum" had a statistically significant influence on science teachers' activity of generating new ideas in the cases of Lithuania and Sweden (Tab. 3.6.14).

**Tab. 3.6.14:** Science teachers' professional development course contents that had a statistically significant effect: based on the data for Lithuania, Sweden, and Singapore

| Innovative activity | Singapore | Lithuania | Sweden | Total |
|---|---|---|---|---|
| Trying out new ideas | Not found | Science curriculum Improving students' critical thinking or inquiry skills | Science curriculum Integrating ICT | Science curriculum |

Analyzing the results from the data on Lithuania, we discovered that the critical thinking and curriculum components of professional development had a statistically significant influence on science teachers' activity of trying out new ideas. However, the professional development of such content has no statistically significant effect on the activity of trying out new ideas in the cases of Singapore and Sweden. The professional development component of integrating information technology into science has a statistically significant influence on trying out new ideas only in the case of Sweden.

*The role of professional development, professional and demographic factors in championing new ideas in science education*

According to De Jong and den Hartog (2010), good idea generators approach problems or performance gaps in professional work activity. "Most ideas need to be promoted as they often do not match what is already used in their work group or organization" (De Jong & den Hartog, 2019, p. 24).

A further implementation of innovation depends on what happens at the championing stage: "full use of an innovation as the best course of action available" or rejection of innovation, that is, "not to adopt an innovation" (Rogers, 2003, p. 177). Scholars state that championing includes finding support and building enthusiasm and confidence about the success of innovation (Howell, Shea, & Higgins, 2005). In this step of innovation process, the degree of uncertainty rests on new ideas, and social reinforcement from other colleagues is needed. To reduce the level of uncertainty, teachers visit another classroom to learn more about teaching, seeking to see how innovation works. Sherry (1997) states that "While information about a new innovation is usually available from outside experts and scientific evaluations, teachers usually seek it from trusted friends and colleagues whose subjective opinions of a new innovation are most convincing" (Sherry, 1997, p. 70).

An OLR analysis to investigate new idea generation activity of science teachers was conducted. We analyzed the influence of different factors on the new idea championing by science teachers—model CHAMP (2). The predictor variables were tested a priori to verify there was no violation of the assumption of multicollinearity.

We tested the hypothesis **H2:** The science teachers' new ideas championing activity is influenced by duration of professional development, gender of teacher, years teacher has been teaching, and content of professional development (2).

We performed the model (CHAMP)–data fit test on the basis of data for the three countries Lithuania, Sweden, and Singapore (Tables 3.6.15 and 3.6.16).

**Tab. 3.6.15:** Model (2) fitting information from an ordinal regression analysis: to explain science teachers' activity of championing new ideas in science education

| Country | Dependent variable | Model | −2 Log Likelihood | $\chi^2$ | df | Sig. | Pearson's $\chi^2$ statistic |
|---|---|---|---|---|---|---|---|
| Lithuania | CHAMP | Intercept only | 1473.368 | | | | |
| | | Final | 1428.983 | 44.385 | 10 | .000 | .067 |
| Sweden | CHAMP | Intercept only | 1215.450 | | | | |
| | | Final | 1145.099 | 70.351 | 10 | .000 | .071 |
| Singapore | CHAMP | Intercept only | 606.157 | | | | |
| | | Final | 587.899 | 18.258 | 10 | .041 | .077 |

**Tab. 3.6.16:** Test of parallel lines: to explain science teachers' activity of championing new ideas in science education

| Country | Model | | −2 Log Likelihood | $\chi^2$ | df | Sig. |
|---|---|---|---|---|---|---|
| Lithuania | CHAMP | Null hypothesis | 1428.983 | | | |
| | | General | 1405.534 | 23.449 | 10 | .059 |
| Sweden | CHAMP | Null hypothesis | 1145.099 | | | |
| | | General | 1086.238 | 58.860 | 10 | .077 |
| Singapore | CHAMP | Null hypothesis | 587.899 | | | |
| | | General | 568.999 | 18.900 | 10 | .052 |

We performed an OLR analysis of the model (CHAMP) on the basis of TIMSS 2015 data for Lithuania. The coefficients of the ordinal regression model based on Wald statistics on the data from Lithuania (Tab. 3.6.17) revealed four statistically significant predictors: gender (GEN), professional experience (PE), duration of professional development (DPD), and professional training content (BTBS 23B—science pedagogy). The positive sign of predictors (GEN, PE, and BTBS 23B) indicates that the predictors have positive effects on science teachers' activity of championing new ideas in education.

In the case of the predictor, gender (GEN), the Wald coefficient had a statistical value of 8.342 and a $p$ value of .04. It means that the probability of championing new ideas in science education of science teachers' becoming statistically significant depends on the teachers' gender. For female Lithuanian science teachers, the odds of championing new ideas were 1.807 (95% CI: 1.188–2.747) times that of male science teachers: $\chi^2(1) = 4.863, p = 0.027$. This role of gender predictor may have been due to big differences between the numbers of male and female teachers in the sample (Tab. 3.6.6).

For the predictor CPD, the Wald coefficient has only one statistical value once, that is, for science pedagogy (BTBS 23B; $p = 0.013 < 0.05$; Tab. 3.6.17). For those Lithuanian science teachers who had science pedagogy as PD content, the odds of championing new ideas were 1.439 (95% CI: 1.076–1.923) times that of those who did not have this PD content, $\chi^2(1) = 6.218, p = 0.013$.

Tab. 3.6.17: Parameter estimates to explain science teachers' activity of championing new ideas in science education: Lithuania

|  |  | Estimate | Std. Error | Wald | df | Sig. | 95% Confidence Interval | |
|---|---|---|---|---|---|---|---|---|
|  |  |  |  |  |  |  | Lower Bound | Upper Bound |
| Threshold | [championing = 1.00] | .829 | .497 | 2.780 | 1 | .004 | -.146 | 1.804 |
|  | [championing = 2.00] | 4.380 | .527 | 69.055 | 1 | .000 | 3.347 | 5.413 |
| Location | GEN | -.014 | .006 | 4.863 | 1 | .027 | -.027 | -.002 |
|  | PE | -.625 | .216 | 8.342 | 1 | .004 | .201 | 1.049 |
|  | BTBS 23A | .293 | .160 | 3.355 | 1 | .067 | -.020 | .606 |
|  | BTBS 23B | .373 | .149 | 6.218 | 1 | .013 | .080 | .665 |
|  | BTBS 23C | .230 | .154 | 2.250 | 1 | .134 | -.071 | .532 |
|  | BTBS 23D | -.119 | .146 | .671 | 1 | .413 | -.405 | .166 |
|  | BTBS 23E | .094 | .149 | .401 | 1 | .526 | -.197 | .385 |
|  | BTBS 23F | .129 | .153 | .721 | 1 | .396 | -.169 | .428 |
|  | BTBS 23G | .272 | .145 | 3.498 | 1 | .061 | -.013 | .556 |
|  | [DPD = 1.00] | -.342 | .154 | 4.925 | 1 | .026 | -.644 | -.040 |
|  | [DPD = 2.00] | 0[a] | . | . | 0 | . | . | . |

Link function: Logit.

The mathematical model for the logit functions of the dependent variable is a part of the CHAMP model created based on data for Lithuania:

$$\ln\frac{P(CHAMP \leq i)}{P(CHAMP > i)} = \begin{cases} .829, \text{ when } i=1 \\ 4.380, \text{ when, } i=2 \end{cases} - 0.014(\text{GEN}) - 0.625(\text{PE})$$

$$+0.373(\text{BTBS23B}) - DPD\begin{cases} .342, \text{ when } i=1 \\ 0, \text{ when, } i=2 \end{cases} \quad (2a)$$

The duration of professional development (DPD) was the categorical variable in our research. It is assumed that duration = 2 is the main category (its coefficient is not present), and we observed how things will change if duration = 1. We see that the coefficient at duration = 1 is negative (−.342). A negative result was obtained for the variable *coding system* in the TIMSS 2015 study: Higher ranks were marked with lower numbers. Since these science teachers spend 16 or more hours of professional development in two years, there is a higher likelihood for those teachers to champion new ideas in science education. For those Lithuanian science teachers who had professional development for 15 hours and less, the odds of championing new ideas were 0.702 (95% CI: 0.521–0.947) times that of those who underwent PD for 16 hours or more: $\chi^2(1) = 4.925, p = 0.026$. An odds ratio < 1 suggests a decreasing probability of dependent variable with increasing values on an independent variable.

The results of our ordinal regression analysis of Lithuania data revealed the importance of science pedagogy in professional development in seeking to increase the activity of championing new ideas in science education by collaboration with colleagues (Tab. 3.6.17). Wald statistics on the data from Lithuania revealed small but a statistically significant negative influence of the number of years the teacher has been teaching (PE) on science teachers' activity of championing new ideas (Tab. 3.6.17). We performed the cross-tabulation in order to determine the influence of the number of years the teacher has been teaching on the new ideas championing activity (Tab. 3.6.18). We found a reverse trend in the number of years the teacher has been teaching and the championing of new ideas (Tab. 3.6.18). This result is due to the data coding order in the TIMSS study. The higher ranks of the CHAMP question were numbered in lower numbers. We compared three groups in the cross-tabulation table (Tab. 3.6.18). The percentage of science teachers in these groups is not similar. It was found that 46.7% of science teachers have 16–30 years of teaching experience. They most often champion new ideas in science education. It is noteworthy that teachers with extensive teaching experience (more than 30 years) also frequently defend new ideas in science education (Tab. 3.6.18). In summary, it can be stated that championing of new ideas increases with increasing number of teaching years (even beyond 30 years).

Tab. 3.6.18: Cross-tabulation of variables based on the data for Lithuania: championing of new ideas and the number of years the teacher has been teaching

|  | Teaching years | | | |
| --- | --- | --- | --- | --- |
| Ranks | 1–15 years | 16–30 years | More 30 years | Total |
| Often | 19.2% | **51.1%** | 29.7% | 100.0% |
| Sometimes | 19.6% | **45.5%** | 34.9% | 100.0% |
| Never | **40.4%** | 43.8% | 15.7% | 100.0% |
| Total | 21.4% | 46.7% | 31.8% | 100.0% |

The predictor variable science pedagogy (ordinal scale) in the OLR analysis was found to contribute to the CHAMP model based on the data for Lithuania. We found a statistically significant influence of the science pedagogy component of professional development on science teachers' activity of championing new ideas (Tab. 3.6.17). The influence of science pedagogy on new ideas championing is statistically significant. The proportional odds model shows a positive effect $\tilde{\beta} = .373$ which is statistically significant according to Wald test with $p = 0.013$. Thus, for science pedagogy, we would say that for a one-unit increase in science pedagogy (i.e., going from 0 to 1), we can expect a 0.373 increase in the ordered log odds of being at a higher level of new ideas championing, with all other variables in the model held constant (Formula 2a).

The ordinal logistic analysis of data for Sweden revealed that four predictor variables (years teaching—TE, BTBS 23C—science curriculum, BTBS 23D—integrating information technology into science, BTBS 23E—improving students' critical thinking or inquiry skills) contribute to the CHAMP model (Tab. 3.6.19). We found a statistically significant positive influence of the science curriculum, integrating information technology into science, and improving students' critical thinking or inquiry skills components of professional development on science teachers' activity of championing new ideas (Table 3.6.19). In the Parameter Estimates table (Table 3.6.19) we can see the coefficients, their standard errors, the Wald test and associated $p$ values (Sig.), and the 95% confidence interval of coefficients in relation to the four predictors are statistically significant.

**Tab. 3.6.19:** Parameter estimates to explain science teachers' activity of championing new ideas in education: Sweden

|  |  | Estimate | Std. Error | Wald | df | Sig. | 95% Confidence Interval | |
|---|---|---|---|---|---|---|---|---|
|  |  |  |  |  |  |  | Lower Bound | Upper Bound |
| Threshold | [CHAMP = 1.00] | 1.999 | .523 | 14.591 | 1 | .000 | .973 | 3.024 |
|  | [CHAMP = 2.00] | 3.949 | .543 | 52.853 | 1 | .000 | 2.885 | 5.014 |
| Location | PE | -.032 | .009 | 12.882 | 1 | .000 | -.049 | -.014 |
|  | GEN | .162 | .160 | 1.032 | 1 | .310 | -.151 | .475 |
|  | BTBS 23A | .197 | .204 | .931 | 1 | .335 | -.203 | .597 |
|  | BTBS 23B | .313 | .230 | 1.851 | 1 | .174 | -.138 | .763 |
|  | BTBS 23C | .458 | .199 | 5.328 | 1 | .021 | .069 | .848 |
|  | BTBS 23D | .717 | .198 | 13.117 | 1 | .000 | .329 | 1.106 |
|  | BTBS 23E | .447 | .212 | 4.456 | 1 | .035 | .032 | .861 |
|  | BTBS 23F | .027 | .188 | .020 | 1 | .888 | -.342 | .395 |
|  | BTBS 23G | -.141 | .185 | .579 | 1 | .447 | -.503 | .221 |
|  | [DPD = 1.00] | -.174 | .258 | .455 | 1 | .500 | -.680 | .332 |
|  | [DPD = 2.00] | 0ª | . | . | 0 | . | . | . |

Link function: Logit.

$$\ln\frac{P(CHAMP \le i)}{P(CHAMP > i)} = \begin{cases} 1.999, \text{when } i = 1 \\ 3.949, \text{when, } i = 2 \end{cases} - 0.032(PE) + 0.458(BTBS23C) \\ + 0.717(BTBS23D) + 0.447(BTBS23E) \quad (2b)$$

According to the Wald test, professional development in the field of integrating information technology into science has the strongest effect on science teachers' activity of championing new ideas $(\hat{\beta} = .717, p = .000)$. The results show a positive effect of science curriculum ($\hat{\beta} = 0.458, p = 0.021$) and improving students' critical thinking or inquiry skills ($\hat{\beta} = 0.447, p = 0.035$; Tab. 3.6.19). For the science teachers from Sweden, who had integrating information technology into science as PD content, the odds of championing new ideas were 1.052 (95% CI: 0.702–1.573) times that of those who did not have this PD content: $\chi^2(1) = 13.117$, $p = 0.000$. For those science teachers who had science curriculum as PD content, the odds of championing new ideas were 1.310 (95% CI: 0.869–1.975) times that of those who did not have this PD content, $\chi^2(1) = 5.328$, $p = 0.021$. For those science teachers who had improving students' critical thinking or inquiry skills as PD content, the odds of championing new ideas were 1.311 (95% CI: 0.702–1.573) times that of those who did not have this PD content, $\chi^2(1) = 4.456$, $p = 0.035$.

We revealed one negative effect of the predictor "number of years the teacher has been teaching" (PE) on Swedish science teachers' activity of championing new ideas in education ($\hat{\beta} = -0.032$, $p = 0.000$) due to the data coding order in the TIMSS study. The higher ranks of the CHAMP model questionnaire were numbered in lower numbers. We performed the cross-tabulation in order to determine the influence of the item "the number of years the teacher has been teaching" on new ideas championing activity (Tab. 3.6.20). It is noteworthy that teachers with extensive teaching experience (16 years or more) are more likely to champion new ideas in science education.

Tab. 3.6.20: Cross-tabulation of variables based on the data for Sweden: championing of new ideas and the number of years the teacher has been teaching

| Ranks | Teaching years | | |
|---|---|---|---|
| | 1–15 years | 16–30 years | More than 30 years |
| Very often | 1.1% | 2.1% | .5% |
| Often | 8.3% | **16.4%** | .5% |
| Sometimes | 37.1% | 48.6% | **55.0%** |
| Never | 53.6% | 32.9% | 44.0% |

We performed an OLR analysis of the TIMSS 2015 data for Singapore (Tab. 3.6.21). It was found that the predictor variable—content of professional development (CPD)—twice statistically significantly contributed to the CHAMP model. We found a statistically significant influence of the science curriculum component of professional development on science teachers' activity championing new ideas ($\hat{\beta} = .637$, $p = .026$; Tab. 3.6.23). The influence of the component of professional development, "addressing individual students' needs," on the science teachers' activity of championing new ideas also statistically significantly contributed to the model ($\hat{\beta} = .432$, $p = .048$; Tab. 3.6.21).

In the case of Singapore, the formula for the explanation of the activity of science teachers championing new ideas in science education (Formula 2c) differs from formulas obtained based on the TIMSS 2015 data for Lithuania (Formula 2a) and for Sweden (Formula 2b).

Professional development and innovative work behavior

However, there are predictors in the formulas that are repeated. The CPD predictor is in all formulas, the predictor PE is in formula (2a) and formula (2b), and the predictor GEN is in formula (2a). The predictors PE and GEN may be conditioned by different teacher samples from Lithuania, Sweden, and Singapore (Tab. 3.6.6).

$$\ln\frac{P(CHAMP \leq i)}{P(CHAMP > i)} = \begin{cases} .373, when\, i = 1 \\ 2.554, when,\, i = 2 \end{cases} + 0.637\big(BTBS23C\big) + 0.432\big(BTBS23G\big)$$

(2c)

Tab. 3.6.21: Parameter estimates to explain science teachers' activity of championing new ideas in education: Singapore

|  |  | Estimate | Std. Error | Wald | df | Sig. | 95% Confidence Interval | |
|---|---|---|---|---|---|---|---|---|
|  |  |  |  |  |  |  | Lower Bound | Upper Bound |
| Threshold | [CHAMP = 1.00] | .373 | .682 | .300 | 1 | .040 | -1.710 | .964 |
|  | [CHAMP = 2.00] | 2.554 | .702 | 13.254 | 1 | .000 | 1.179 | 3.929 |
| Location | BTBS 23A | -.047 | .275 | .029 | 1 | .864 | -.586 | .491 |
|  | BTBS 23B | -.286 | .412 | .480 | 1 | .488 | -1.094 | .522 |
|  | BTBS 23C | .637 | .286 | 4.965 | 1 | .026 | .077 | 1.197 |
|  | BTBS 23D | -.078 | .248 | .100 | 1 | .752 | -.564 | .407 |
|  | BTBS 23E | -.263 | .270 | .949 | 1 | .330 | -.793 | .266 |
|  | BTBS 23F | .353 | .250 | 1.989 | 1 | .158 | -.137 | .843 |
|  | BTBS 23G | .432 | .245 | 3.108 | 1 | .048 | -.048 | .913 |
|  | GEN | -.206 | .230 | .801 | 1 | .371 | -.657 | .245 |
|  | PE | .012 | .013 | .793 | 1 | .373 | -.014 | .038 |
|  | [DPD = 1.00] | .147 | .242 | .370 | 1 | .543 | -.327 | .621 |
|  | [DPD = 2.00] | 0[a] | . | . | 0 | . | . | . |

Link function: Logit.

For those science teachers from Singapore who had science curriculum as PD content, the odds of sharing experiences were 1.765 (95% CI: 1.003–3.105) times that of those who did not have this PD content, $\chi^2(1) = 4.965$, $p = 0.02$. For those Singapore science teachers whose PD content included addressing individual students' needs, the odds of sharing experiences were 1.541 (95% CI: 0.954–2.511) times that of those who did not have this PD content, $\chi^2(1) = 3.108$, $p = 0.048$.

We compared the results of the influence of predictor CPD on science teachers' new ideas championing activity in different countries (Tab. 3.6.22). Summarizing the results of our OLR analysis of data for Lithuania, Sweden, and Singapore about the influence of CPD on science teachers' activity of championing new ideas, we noticed that the CPD component science curriculum had a twice statistically significant influence on science teachers' activity of championing new ideas; this was particularly the case for data on Sweden and Singapore (Tab. 3.6.22).

Tab. 3.6.22: The content of science teachers' professional development that had a statistically significant effect on their championing new ideas: based on the data for Lithuania, Sweden, and Singapore

| Innovative activity | Singapore | Lithuania | Sweden |
|---|---|---|---|
| Championing new ideas | Science curriculum Individual needs | Science pedagogy | Science curriculum Integrating ITC Improving students' critical thinking or inquiry skills |

Analyzing the results drawn from the data for Singapore, we found a statistically significant influence of the CPD item "addressing individual students' needs" on the science teachers' activity of championing new ideas. However, the same item did not have a statistically significant effect on the activity of generating new ideas for teachers of Lithuania and Sweden. The professional development items "integrating information technology into science" and "improving students' critical thinking or inquiry skills" had a statistically significant influence on championing new ideas only for teachers from Sweden. Summarizing the results of a comparative analysis on the influence of CPD on science teachers' activity of championing new ideas in science classroom, we noticed that the responses to the CPD item "science curriculum" were most often repeated.

*The role of professional development, professional and demographic factors in the implementation of new ideas in science education*

Layton (1986) stated that innovations has become a permanent feature of science education not only in curriculum content but also in the associated teaching methods and materials. "Innovation in science education is less a characteristic of a particular period in time than normal and continuing process. The rapid advance of scientific knowledge and the emergence of significant technologies alone require that this is so [...] Even so, the events of the past thirty years provide examples of planned innovation on a scale rarely witnessed previously" (Layton, 1986, p. 9).

The last three decades have seen growing calls for innovations associated with teaching methods. In light of science education reforms (NRC, 1996, 2000, 2012), teaching methods have emerged that focused on innovations that have alternately been called scientific inquiry, discovery, and constructivist approaches (Furtak, & Kunter, 2012). "These instructional approaches feature the teacher in the role of facilitator of student learning, providing varying levels of instructional support for students as they engage in the thinking processes and activities of scientists to develop conceptual understandings of the big ideas in science" (Furtak, & Kunter, 2012, p. 284).

Schools must therefore see that learners develop generic abilities: usage of information, creating solutions for problems, collaboration with others, and strong communication skills. These new societal needs encourage better professional development of preservice teachers, including the development of their problem-solving abilities, control of information, communication, and collaboration. Every teacher should be competent to help young people develop these generic abilities; therefore, these abilities need to be developed during the preservice stage of teachers' education. This is what is missing in educational practice: Many scientists teach concepts, principles, and formulas regarding course subjects, and then they conventionally solve several sample problems. After they complete the instruction, learners are usually asked whether or not they have understood the concepts, principles, and so on. Even though most learners claim they understand fundamental principle(s)/concept(s), they are not able to solve concept-related problems. It seems learners have no problem-solving abilities. They cannot develop any systematic problem-solving strategies in this way. As a result, the activities or performance of the learners do not reflect their success (Gok, 2014).

We tested the hypothesis **H3:** The promoting of problem-solving activity in science classroom is influenced by the duration of professional development, gender of the teacher, number of years the teacher has been teaching, and content of professional development (3).

We tested the PPS model using the TIMSS 2015 (model 3) data for the three countries—Lithuania, Sweden, and Singapore.

## *The results of promotion of the problem-solving (PPS)*

We compare a model without any explanatory variables (the baseline or "Intercept Only" model) with the model with all the explanatory variables (the "Final" model—this would normally have several explanatory variables, but we have used only the *level of education* variable from the TIMSS 2015 database; Tab. 3.6.23). The significant $\chi^2$ ($p < 0.05$) indicates that the final model gives a significant improvement over the baseline intercept-only model. The small $p$ value from the test for the model–data fit, $p < 0.05$, allows us to conclude that at least one of the regression coefficients in the model is not equal to zero. If we do not reject this hypothesis' Pearson's $\chi^2$ statistic for the model, that is, $p > .05$, then we can conclude that the data and the model predictions are similar and that we have a good model.

Tab. 3.6.23: Model (PPS) fitting information from an ordinal regression analysis

| Country | Dependent variable | Model | −2 Log Likelihood | $\chi^2$ | df | Sig. |
|---|---|---|---|---|---|---|
| Lithuania | PPS | Intercept only | 1909.681 | | | |
| | | Final | 1877.300 | 32.381 | 10 | .000 |
| Sweden | PPS | Intercept only | 995.922 | | | |
| | | Final | 974.067 | 21.856 | 10 | .016 |
| Singapore | PPS | Intercept only | 466.319 | | | |
| | | Final | 437.576 | 28.743 | 10 | .001 |

Wald statistics for the data of Lithuania revealed two statistically significant predictors from the CPD group: science pedagogy (BTBS 23B; $p = 0.049 < 0.05$) and improving students' critical thinking or inquiry skills (BTBS 23E; $p = 0.001 < 0.05$; Tab. 3.6.24). For those Lithuanian science teachers who had science pedagogy as PD content, the odds of championing new ideas were 1.256 (95% CI: 0.965–1.637) times that of those who did not have this PD content, $\chi^2(1) = 0.375$, $p = 0.011$. For those Lithuanian science teachers who had improving students' critical thinking or inquiry skills as PD content, the odds of championing new ideas were 1.590 (95% CI: 1.223–2.068) times that of those who did not have this PD content, $\chi^2(1) = 0.448, p = 0.001$.

We found a statistically significant influence of the predictor PE (professional experience) on the item "science teachers' activity of promoting problem-solving in the science classroom" ($p = 0.003 < 0.05$) which we selected from the data for Lithuania (Tab. 3.6.24). For those Lithuanian science teachers who had more professional experience, the odds of championing new ideas were 0.984 (95% CI: 972–0.995) times that of those who had less professional experience, $\chi^2(1) = -0.017, p = 0.003$. An odds ratio < 1 suggests a decreasing probability of the dependent variable with increasing values on an independent variable. The estimated values of Wald statistics show a negative influence of the predictor PE on science teachers' activity of promoting problem-solving in the science classroom; this occurred because of data coding order in the TIMSS study. The higher ranks of the PPS question were numbered in lower numbers.

Tab. 3.6.24: Parameter estimates to explain science teachers promoting problem-solving activity in science education: Lithuania

|  |  | Estimate | Std. Error | Wald | df | Sig. | 95% Confidence Interval | |
|---|---|---|---|---|---|---|---|---|
|  |  |  |  |  |  |  | Lower Bound | Upper Bound |
| Threshold | [PPS = 1] | -1.417 | .456 | 9.658 | 1 | .002 | -2.311 | -.523 |
|  | [PPS = 2] | .352 | .452 | .605 | 1 | .037 | -.535 | 1.238 |
|  | [PPS = 3] | 4.002 | .489 | 66.990 | 1 | .000 | 3.044 | 4.960 |
| Location | PE | -.017 | .006 | 8.752 | 1 | .003 | -.029 | -.006 |
|  | GEN | -.129 | .192 | .456 | 1 | .500 | -.505 | .246 |
|  | BTBS 23A | -.141 | .144 | .959 | 1 | .327 | -.423 | .141 |
|  | BTBS 23B | .265 | .135 | 3.846 | 1 | .049 | .000 | .531 |
|  | BTBS 23C | .166 | .140 | 1.418 | 1 | .234 | -.107 | .440 |
|  | BTBS 23D | -.130 | .133 | .953 | 1 | .329 | -.391 | .131 |
|  | BTBS 23E | .448 | .136 | 10.908 | 1 | .001 | .182 | .713 |
|  | BTBS 23F | .014 | .139 | .010 | 1 | .920 | -.258 | .286 |
|  | BTBS 23G | .098 | .132 | .549 | 1 | .459 | -.161 | .356 |
|  | [DPD = 1.00] | -.058 | .141 | .168 | 1 | .682 | -.334 | .218 |
|  | [DPD = 2.00] | 0[a] | . | . | 0 | . | . | . |

Link function: Logit.
a. This parameter is set to zero because it is redundant.

We compared three groups in the cross-tabulation table (Tab. 3.6.25). The percentage of science teachers in these groups is not similar. It was found that 46.4% of science teachers have 16–30 years of teaching experience. They often promote problem-solving in science education.

Wald coefficients for the data for Sweden indicate a statistically significant association between science teachers' new idea sharing activity and the number of years the teacher has been teaching (PE; $p = 0.001 < 0.05$). The estimate parameter has a negative sign. The negative sign indicates a reverse association between science teachers' activity of promoting problem-solving in the science classroom and the number of years they have been teaching (PE). This result is due to the data coding order in the TIMSS 2015 study. The higher ranks of PPS question were numbered in lower numbers. We performed data cross-tabulation. The cross-tabulation of variables did not confirm this tendency (Tab. 3.6.25).

Tab. 3.6.25: Cross-tabulation of the variables "promoting problem-solving" and "the number of years the teacher has been teaching" (professional experience [PE]): Lithuania

| The number of years the teacher has been teaching | Every or almost every lesson | Ranks of promotion of problem-solving activity | | |
|---|---|---|---|---|
| | | About half the lessons | Some lessons | Never |
| From 1 to 15 years | 14.4% | 21.2% | 22.7% | **33.3%** |
| From 16 to 30 years | 39.2% | 46.7% | 48.6% | 46.7% |
| More 30 years | **46.4%** | 32.1% | 28.7% | 20.0% |
| % of total | 100.0% | 100.0% | 100.0% | 100.0% |

The mathematical model of the data from Lithuania for the logarithms of the dependent variable is created based on the PPS model:

$$\ln\frac{P(PPS \leq i)}{P(PPS > i)} = \begin{cases} -1.417, when\, i = 1 \\ .352, when,\, i = 2 \\ 4.002, when\, i = 3 \end{cases} + 0.265(BTBS23B) + 0.448(BTBS23E) - .017(PE)$$

(3a)

The coefficients in the PPS model of the data from Lithuania (1a) have both positive and negative signs. The positive sign indicates predictors BTBS 23B and BTBS 23E have positive effects on the item "promoting problem-solving in science education."

## Sweden

We get a statistically significant confirmation that the probability of science teachers' sharing of new ideas in science education of teachers' depends on the content of professional development (Tab. 3.6.26). For the predictor CPD, the Wald coefficient has a statistically significant value for items: science curriculum training (BTBS 23C; $p = 0.006 < 0.05$) and science pedagogy (BTBS 23A; $p = 0.026 < 0.05$; Tab. 3.6.26). For those science teachers from Sweden who had science pedagogy as PD content, the odds of sharing experiences were 1.002 (95% CI: 0.985–1.019) times that of those who did not have this PD content, $\chi^2(1) = 0.444, p = 0.026$. For the science teachers from Sweden who had science curriculum as PD content, the odds of sharing experiences were 2.470 (95% CI: 1.650–3.714) times that of those who did not have this PD content, $\chi^2(1) = 0.544, p = 0.006$.

Tab. 3.6.26: Parameter estimates to explain science teachers promoting problem-solving activity in science education: Sweden

|  |  | Estimate | Std. Error | Wald | df | Sig. | 95% Confidence Interval | |
|---|---|---|---|---|---|---|---|---|
|  |  |  |  |  |  |  | Lower Bound | Upper Bound |
| Threshold | [PPS = 1] | 2.810 | .564 | 24.808 | 1 | .000 | 3.916 | 1.704 |
|  | [PPS = 2] | 1.071 | .554 | 3.746 | 1 | .043 | 2.157 | .014 |
| Location | GEN | -.255 | .157 | 2.638 | 1 | .104 | -.563 | .053 |
|  | BTBS 23A | .444 | .200 | 4.928 | 1 | .026 | .837 | .052 |
|  | BTBS 23B | .199 | .234 | .719 | 1 | .396 | -.658 | .260 |
|  | BTBS 23C | .544 | .200 | 7.431 | 1 | .006 | .153 | .936 |
|  | BTBS 23D | .045 | .196 | .052 | 1 | .819 | -.339 | .428 |
|  | BTBS 23E | .452 | .211 | 4.573 | 1 | .052 | .038 | .866 |
|  | BTBS 23F | -.268 | .189 | 2.014 | 1 | .156 | -.638 | .102 |
|  | BTBS 23G | -.134 | .184 | .535 | 1 | .465 | -.495 | .226 |
|  | PE | -.281 | .154 | 3.322 | 1 | .068 | -.582 | .021 |
|  | [duration = 1.00] | -.071 | .260 | .074 | 1 | .786 | -.581 | .439 |
|  | [duration = 2.00] | 0[a] | . | . | 0 | . | . | . |

Link function: Logit.
a. This parameter is set to zero because it is redundant.

The mathematical model for the logit functions of the dependent variable is created based on the PPS model of the data for Sweden:

$$\ln\frac{P(GNI \le i)}{P(GNI > i)} \begin{cases} 2.810, when\, i = 1 \\ 1.071, when,\, i = 2 \end{cases} + 0.444\left(BTBS23A\right) + 0.544\left(BTBS23C\right) \quad (3b)$$

*Singapore*

We tested the PPS model of the data for Singapore. Wald statistics regarding Lithuanian data revealed three statistically significant predictors from the CPD group: science pedagogy (BTBS 23B; $p = 0.026 < 0.05$), science curriculum (BTBS 23C; $p = 0.037 < 0.05$), and addressing individual students' needs (BTBS 23G; $p = 0.008 < 0.05$; Tab. 3.6.27).

For those Singaporean science teachers who had science curriculum as PD content, the odds of sharing experiences were 1.697 (95% CI: 0.969–2.971) times that of those who did not have this PD content, $\chi^2(1) = 4.350$, $p = 0.037$; with science pedagogy as PD content, the odds of sharing experiences were 0.404 (95% CI: 0.193–0.910) times that of those who did not have this PD content, $\chi^2(1) = 4.958$, $p = 0.026$; with individual students' needs as PD content, the odds of sharing experiences were 1.811 (95% CI: 1.136–2.888) times that of those who did not have this PD content, $\chi^2(1) = 7.042$, $p = 0.008$.

Tab. 3.6.27: Parameter estimates to explain science teachers promoting problem-solving activity in science education: Singapore

| Parameter Estimates | | | | | | | | |
|---|---|---|---|---|---|---|---|---|
| | | Estimate | Std. Error | Wald | df | Sig. | 95% Confidence Interval | |
| | | | | | | | Lower Bound | Upper Bound |
| Threshold | [PPS = 1.00] | .872 | .749 | 1.355 | 1 | .244 | -.596 | 2.339 |
| | [PPS = 2.00] | 3.692 | .780 | 22.384 | 1 | .000 | 2.163 | 5.222 |
| Location | BTBS 23A | .290 | .272 | 1.137 | 1 | .286 | -.243 | .824 |
| | BTBS 23B | .927 | .416 | 4.958 | 1 | .026 | 1.743 | .111 |
| | BTBS 23C | .589 | .283 | 4.350 | 1 | .037 | .036 | 1.143 |
| | BTBS 23D | .124 | .244 | .257 | 1 | .612 | .603 | .355 |
| | BTBS 23E | .204 | .268 | .579 | 1 | .447 | -.322 | .730 |
| | BTBS 23F | .245 | .245 | .999 | 1 | .318 | -.236 | .726 |

Tab. 3.6.27: Continued

| Parameter Estimates | | | | | | | | |
|---|---|---|---|---|---|---|---|---|
| BTBS 23G | .643 | .242 | 7.042 | 1 | .008 | .168 | 1.118 |
| GEN | .215 | .227 | .898 | 1 | .343 | -.230 | .660 |
| PE | -.159 | .221 | .516 | 1 | .473 | -.593 | .275 |
| [DPD = 1.00] | -.149 | .239 | .387 | 1 | .534 | -.616 | .319 |
| [DPD = 2.00] | 0ª | . | . | 0 | . | . | . |

Link function: Logit.
a. This parameter is set to zero because it is redundant.

The mathematical model for the logit functions of the dependent variable PPS is created based on the data for Singapore:

$$\ln\frac{P(GNI \le i)}{P(GNI > i)} = \begin{cases} .872, \text{ when } i=1 \\ 3.692, \text{ when, } i=2 \end{cases} + 0.927(BTBS23B) + 0.589(BTBS23C) + 0.643(BTBS23G) \quad (3c)$$

We performed the "test of parallel lines." The test results confirm the null hypothesis that the PPS model has one set of coefficients for the data for Lithuania, Sweden, and Singapore (Tab. 3.6.28).

Tab. 3.6.28: Test of parallel lines to explain the PPS model: Lithuania, Sweden, and Singapore

| Country | Model | | -2 Log Likelihood | $\chi^2$ | df | Sig. |
|---|---|---|---|---|---|---|
| Lithuania | PPS | Null hypothesis | 1285.029 | | | |
| | | General | 1271.627 | 13.402 | 10 | .202 |
| Sweden | PPS | Null hypothesis | 974.067 | | | |
| | | General | 926.885a | 47.182b | 30 | .054 |
| Singapore | PPS | Null hypothesis | 437.576 | | | |
| | | General | 428.080 | 9.496 | 10 | .486 |

The comparative OLR analysis of results based on the data for the three countries (Lithuania, Sweden, and Singapore) revealed the main predictors of $CPD_i$ for science teachers' ability to share new ideas (Tab. 3.6.29). Summarizing the results of our OLR analysis of the data for Lithuania, Sweden, and Singapore in relation to the influence of $CPD_i$ (Content of professional development variables) predictor on science teachers' activity of promoting problem-solving, we found that the CPD topics science curriculum and science pedagogy had a twice statistically significant association with the activity of promoting problem-solving (Tab. 3.6.29). We found that professional development items "integrating information and communication technology in science" and "science assessment" were not associated with science teachers' problem-solving activity; we found this to be a common feature for the data for all the three countries—Lithuania, Sweden, and Singapore (Tab. 3.6.29). Gender (GEN) and duration of professional development (DPD) were not associated statistically significantly with the science teachers' promotion of problem-solving activity; this again was a common feature in our analysis of the data for the three countries. The professional experience of science teachers is statistically significantly associated with promotion of problem solving activity in science education only for the data of Lithuania.

Sharing of new ideas occurs by communication and thus by communications channels (Rogers, 2003). Communication is central to the interaction within a social context. Communication is central to collaborative work as well as to the interactions taking place within a social context. The teamwork approach offers a vehicle for the transfer of the control of learning from teachers to learners, encourages students' responsibility for learning, and promotes intrinsic motivation (Lawlor et al., 2015). Individual team members enjoy the support of their peers and shared responsibility (Pyle, 1995). This opportunity enables them to avoid feeling helpless and suffering from poor motivation in formal learning contexts (Pell et al., 2007). Innovative activities with various societal actors in formal and informal scenarios open up new opportunities for science education within and beyond schools (Okada, 2013).

**Tab. 3.6.29:** The content of science teachers' professional development that had a statistically significant influence on science teacher's promoting problem-solving in education: Lithuania, Sweden, and Singapore

| Question code ($CPD_i$) | Content of professional development | Lithuania | Sweden | Singapore |
|---|---|---|---|---|
| BTBS 23A | Science content |  | + |  |
| BTBS 23B | Science pedagogy | + |  | + |
| BTBS 23C | Science curriculum |  | + | + |
| BTBS 23D | Integrating information technology into science |  |  |  |
| BTBS 23E | Improving students' critical thinking or inquiry skills | + |  |  |
| BTBS 23F | Science assessment |  |  |  |
| BTBS 23G | Addressing individual students' needs |  |  | + |

## The role of professional development, professional and demographic factors in sharing new ideas in science education

We tested the hypothesis H4: Science teachers' activity of sharing new ideas is influenced by duration of professional development, gender of the teacher, years the teacher has been teaching, and content of professional development (4).

We tested the SHARE model in the TIMSS 2015 database for the three countries Lithuania, Sweden, and Singapore. We compare a model without any explanatory variables (the baseline or "Intercept Only" model) with the model with all the explanatory variables (the "Final" model—this would normally have several explanatory variables but the one we used contains only the level of education; Tab. 3.6.30). The significant $\chi^2$ statistic ($p < 0.05$) indicates that the final model gives a significant improvement over the baseline intercept-only model. The small $p$ value from the test for the model–data fit, $p < 0.05$, allows to conclude that at least one of the regression coefficients in the model is not equal to zero. If we do not reject this hypothesis, and the Pearson's $\chi^2$ statistic for the model ($p > 0.05$), then we can conclude that the data and the model predictions are similar and that we have a good model (Tab. 3.6.30).

Tab. 3.6.30: The SHARE model fitting information from an ordinal regression analysis: Lithuania, Sweden, and Singapore

| Country | Dependent variable | Model | −2 Log Likelihood | $\chi^2$ | df | Sig. | Pearson's $\chi^2$ statistic |
|---|---|---|---|---|---|---|---|
| Lithuania | SHARE | Intercept only | 1313.286 | | | | |
| | | Final | 1285.029 | 28.257 | 10 | .002 | .067 |
| Sweden | SHARE | Intercept only | 1264.079 | | | | |
| | | Final | 1183.386 | 80.694 | 12 | .000 | .122 |
| Singapore | SHARE | Intercept only | 556.826 | | | | |
| | | Final | 537.109 | 19.716 | 10 | .032 | .692 |

Wald statistics on the data for Lithuania revealed two statistically significant predictors for the CPD group: science curriculum (BTBS 23C) and improving students' critical thinking or inquiry skills (Tab. 3.6.31). For the predictor CPD, Wald coefficient has statistical values for two items: science curriculum training (BTBS 23C; $p = 0.011 < 0.05$) and improving students' critical thinking or inquiry skills (BTBS 23E; $p = .017 < .05$; Tab. 3.6.31). For those Lithuanian science teachers who had science curriculum as PD content, the odds of championing new ideas were 1.481 (95% CI: 1.111–1.974) times that of those who did not have this PD content, $\chi^2(1) = 0.375$, p = 0.011; with improving students' critical thinking or inquiry skills as PD content, the odds of championing new ideas were 1.414 (95% CI: 1.067–1.874) times that of those who did not have this PD content, $\chi^2(1) = 0.345$, p = 0.017.

The SHARE model created based on the data for Lithuania:

$$\ln\frac{P(SHARE \leq i)}{P(SHARE > i)} = \begin{cases} 2.312, \text{ when } i=1 \\ 6.327, \text{ when, } i=2 \end{cases} + 0.375(BTBS23C) + 0.345(BTBS23E)$$

(4a)

All coefficients in the SHARE model based on the data for Lithuania (1a) have a positive sign. The positive signs indicate that the predictors BTBS 23C and BTBS 23E have positive effects on science teachers' activity of sharing new ideas.

Tab. 3.6.31.: Parameter estimates for the SHARE model: Lithuania

|  |  | Estimate | Std. Error | Wald | df | Sig. | 95% Confidence Interval | |
|---|---|---|---|---|---|---|---|---|
|  |  |  |  |  |  |  | Lower Bound | Upper Bound |
| Threshold | [SHARE = 1.00] | 2.312 | .490 | 22.297 | 1 | .000 | 1.352 | 3.272 |
|  | [SHARE = 2.00] | 6.327 | .570 | 123.022 | 1 | .000 | 5.209 | 7.445 |
| Location | PE | -.002 | .006 | .086 | 1 | .769 | -.014 | .010 |
|  | GEN | .327 | .201 | 2.637 | 1 | .104 | -.068 | .721 |
|  | BTBS 23A | .168 | .152 | 1.226 | 1 | .268 | -.130 | .467 |
|  | BTBS 23B | .004 | .143 | .001 | 1 | .976 | -.276 | .285 |
|  | BTBS 23C | .375 | .147 | 6.498 | 1 | .011 | .087 | .663 |
|  | BTBS 23D | .233 | .141 | 2.718 | 1 | .099 | -.044 | .510 |
|  | BTBS 23E | .345 | .144 | 5.713 | 1 | .017 | .062 | .627 |
|  | BTBS 23F | -.017 | .147 | .013 | 1 | .908 | -.304 | .270 |
|  | BTBS 23G | .075 | .140 | .286 | 1 | .593 | -.199 | .349 |
|  | [DPD = 1.00] | -.054 | .150 | .129 | 1 | .719 | -.347 | .239 |
|  | [DPD = 2.00] | 0[a] | . | . | 0 | . | . | . |

Link function: Logit.
a. This parameter is set to zero because it is redundant.

Wald coefficients in the data for Sweden indicate a statistically significant association between science teachers' new idea sharing activity and professional experience (PE; $p = 0.001 < 0.05$; Tab. 3.6.32). The estimate parameter has the negative sign because higher ranks of the SHARE question were numbered in lower numbers. The negative sign indicates a reverse association between science teachers' activity of sharing new ideas and the number of years they have been teaching. The cross-tabulation of variables does not confirm this tendency (Tab. 3.6.32). The cross-tabulation data revealed that teachers with 16–30 years of experience tend to share innovative ideas *often* and *very often*.

**Tab. 3.6.32:** Cross-tabulation of variables on the data for Sweden: sharing new ideas and the number of years the teacher has been teaching

| Years been teaching | Ranks of new idea sharing | | | | Total |
|---|---|---|---|---|---|
| | Very Often | Often | Sometimes | Never | |
| From 1 to 15 years | 24.3% | 39.6 | 32.8% | 3.4% | 69.7% |
| From 16 to 30 years | **32.4%** | **44.1** | 18.4% | 5.1% | 21.5% |
| More 30 years | 23.7% | 52.6 | 23.7% | .0% | 8.8% |

We get a statistically significant confirmation that the probability of science teachers' sharing new ideas in science education depends on the content of professional development (Tab. 3.6.33). For the predictor CPD, Wald coefficient has a statistically significant value for three items: science curriculum training (BTBS 23C; $p = 0.034 < 0.05$), integrating information technology into science (BTBS 23D; $p = 0.002 < 0.05$), and improving students' critical thinking or inquiry skills (BTBS 23E; $p = 0.043 < .05$; Tab. 3.6.33).

Tab. 3.6.33: Parameter estimates for the SHARE model: Sweden

|  |  | Estimate | Std. Error | Wald | df | Sig. | 95% Confidence Interval | |
|---|---|---|---|---|---|---|---|---|
|  |  |  |  |  |  |  | Lower Bound | Upper Bound |
| Threshold | [share2 = 1.00] | 1.220 | .585 | 4.346 | 1 | .037 | .073 | 2.367 |
|  | [share2 = 2.00] | 3.196 | .599 | 28.496 | 1 | .000 | 2.023 | 4.370 |
| Location | PE | -.031 | .009 | 11.988 | 1 | .001 | -.048 | -.013 |
|  | GEN | .164 | .161 | 1.028 | 1 | .311 | -.153 | .480 |
|  | BTBS 23A | .004 | .213 | .000 | 1 | .984 | -.414 | .423 |
|  | BTBS 23B | .188 | .235 | .639 | 1 | .424 | -.273 | .649 |
|  | BTBS 23C | .422 | .200 | 4.477 | 1 | .034 | .031 | .813 |
|  | BTBS 23D | .616 | .203 | 9.256 | 1 | .002 | .219 | 1.013 |
|  | BTBS 23E | .430 | .212 | 4.107 | 1 | .043 | .014 | .847 |
|  | BTBS 23F | -.039 | .190 | .043 | 1 | .836 | -.412 | .333 |
|  | BTBS 23G | -.198 | .187 | 1.122 | 1 | .289 | -.565 | .168 |
|  | [DPD = 1.00] | .470 | .330 | 2.032 | 1 | .154 | -.176 | 1.116 |
|  | [DPD = 2.00] | .005 | .301 | .000 | 1 | .986 | -.584 | .594 |
|  | [DPD = 3.00] | -.307 | .270 | 1.289 | 1 | .256 | -.836 | .223 |
|  | [DPD = 4.00] | 0[a] | . | . | 0 | . | . | . |

Link function: Logit.
a. This parameter is set to zero because it is redundant.

The SHARE model for the logit functions based on the data for Sweden:

$$ln\frac{P(SHARE \le i)}{P(SHARE > i)} = \begin{cases} 1.220, \text{ when } i=1 \\ 3.196, \text{ when, } i=2 \end{cases} - 0.031(PE) + 0.422(BTBS23C) \\ + 0.616(BTBS23D) + 0.430(BTBS23E) \quad (4b)$$

All coefficients of CPD in the SHARE mathematical model based on the data for Sweden (4b) have positive signs. The positive sign indicates that three predictors have positive effects on science teachers' activity of sharing new ideas in education (Formula 4b).

We analyzed the data for Singapore to ascertain the influence of different predictors (DPD, CPD, GEN, PE) on science teachers' activity of sharing new ideas according to the SHARE model. The results of Wald statistic revealed that only one predictor (CPD's improving students' critical thinking or inquiry skills) had a statistically significant influence on the dependent variable SHARE (Tab. 3.6.34).

For those Singaporean science teachers who had inquiry skills as PD content, the odds of sharing experiences were 1.386 (95% CI: 0.838–2.293) times that of those who did not have this PD content, $\chi^2(1) = 2.440$, $p = 0.018$.

**Tab. 3.6.34:** Parameter estimates for the SHARE model: Singapore

|  |  | Estimate | Std. Error | Wald | df | Sig. | 95% Confidence Interval | |
|---|---|---|---|---|---|---|---|---|
|  |  |  |  |  |  |  | Lower Bound | Upper Bound |
| Threshold | [sharing = 1.00] | 1.812 | .700 | 6.704 | 1 | .010 | .440 | 3.183 |
|  | [sharing = 2.00] | 5.032 | .772 | 42.432 | 1 | .000 | 3.518 | 6.545 |
| Location | BTBS 23A | -.063 | .277 | .052 | 1 | .819 | -.606 | .479 |
|  | BTBS 23B | -.069 | .411 | .028 | 1 | .866 | -.874 | .736 |
|  | BTBS 23C | .253 | .283 | .800 | 1 | .371 | -.302 | .809 |
|  | BTBS 23D | -.137 | .252 | .293 | 1 | .588 | -.631 | .358 |
|  | BTBS 23E | .422 | .270 | 2.440 | 1 | .018 | -.107 | .951 |
|  | BTBS 23F | .426 | .249 | 2.923 | 1 | .087 | -.062 | .915 |
|  | BTBS 23G | .266 | .249 | 1.140 | 1 | .286 | -.222 | .755 |
|  | BTBG 02 | -.038 | .235 | .027 | 1 | .870 | -.499 | .422 |
|  | BTBG 01 | -.005 | .014 | .119 | 1 | .730 | -.032 | .022 |
|  | [duration = 1.00] | .168 | .245 | .474 | 1 | .491 | -.311 | .648 |
|  | [duration = 2.00] | 0ª | . | . | 0 | . | . | . |

Link function: Logit.
a. This parameter is set to zero because it is redundant.

The mathematical model SHARE showing the logit functions of the dependent variable is created based on the data for Singapore:

$$\ln\frac{P(SHARE \leq i)}{P(SHARE > i)} = \begin{cases} 1.812, when\, i = 1 \\ 5.032, when,\, i = 2 \end{cases} + 0.422(BTBS23E) \qquad (4b)$$

We performed the "test of parallel lines." The test results confirm the null hypothesis that the SHARE mode has one set of coefficients in the data for all the three chosen countries—Lithuania, Sweden, and Singapore (Tab. 3.6.35).

Tab. 3.6.35: Test of parallel lines for the SHARE model: Lithuania, Sweden, and Singapore

| Country | Model | | -2 Log Likelihood | $\chi^2$ | df | Sig. |
|---|---|---|---|---|---|---|
| Lithuania | SHARE | Null hypothesis | 1285.029 | | | |
| | | General | 1271.627 | 13.402 | 10 | .202 |
| Sweden | SHARE | Null hypothesis | 1214.197 | | | |
| | | General | 1193.036 | 11.160 | 12 | .232 |
| Singapore | SHARE | Null hypothesis | 537.109 | | | |
| | | General | 497.622 | 39.487 | 20 | .066 |

### 3.6.4. Discussion

The results of the comparative OLR analysis of data for the three countries (Lithuania, Sweden, and Singapore) revealed the influence of the main predictors of CPD on the science teachers' ability to share new ideas (Tab. 3.6.36). Summarizing the results of our OLR analysis of the data for Lithuania, Sweden, and Singapore in relation to the influence of CPD predictors on science teachers' activity of sharing of new ideas, we noticed that the CPD component "improving students' critical thinking or inquiry skills" had a statistically significant influence on science teachers' activity of sharing new ideas in the data for all the three countries (Lithuania, Sweden, and Singapore; Tab. 3.6.36). We found a statistically significant role of professional development about science curriculum in science teachers' activity of sharing new ideas in education.

Tab. 3.6.36: The content of science teachers' professional development that had a statistically significant effect on science teachers sharing new ideas in education: Lithuania, Sweden, and Singapore

| Question code (CPD$_i$) | Content of professional development | Lithuania | Sweden | Singapore |
|---|---|---|---|---|
| BTBS 23A | Science content | | | |
| BTBS 23B | Science pedagogy | | | |
| BTBS 23C | Science curriculum | + | + | |
| BTBS 23D | Integrating information technology into science | | + | |
| BTBS 23E | Improving students' critical thinking or inquiry skills | + | + | + |
| BTBS 23F | Science assessment | | | |
| BTBS 23G | Addressing individual students' needs | | | |

Analyzing the results of other predictors (DPD, GEN, PE) we found that only one predictor, professional experience (PE), had a statistically significant influence on the new ideas sharing activity of science teachers from Sweden.

We found from the TIMSS 2015 data for Lithuania that among Lithuanian science teachers the duration of professional development (DPD) had an effect on their championing of new ideas. We found that science teachers spend 16 or more hours of professional development in two years, making it more likely that science teachers will better champion new ideas in science education. It means that championing new ideas in science education remains sensitive for the duration of professional development, especially for teachers with greater professional experience. It must be stated that the professional experience (years teaching) of Lithuanian teachers is longer than that of teachers in Singapore and Sweden (Tab. 3.6.6). It would be appropriate for future studies to investigate the associations between duration of professional development, teachers professional experience, and innovative work behavior components. We found that the influence of professional development duration on science teachers' innovative work behavior is more the exception than the rule. In the case of only one country (Lithuania) and only one activity (championing), a statistically significant association was established between the two—professional development duration and innovative work behavior.

We found a statistically significant effect of science teachers' professional experience (PE) on all innovative activities: generating, championing, implementing (problem-solving), and sharing of new ideas. We noticed a statistically significant association between science teachers' professional experience (number of years the teacher has been teaching) on trying out new ideas on the data for Lithuania; between professional experience and championing new ideas on the data for Sweden and Lithuania; between professional experience and implementation of new ideas (problem-solving) on the data for Lithuania; and between professional experience and sharing of new ideas on the data for Sweden. It should be noted that there was no statistically significant association between the components of teachers' professional experience and innovative work behavior (trying out, championing, implementation, and sharing new ideas) on the data for Singapore. This result may have been due to the significantly lower duration of professional experience of Singapore teachers over the years (Tab. 3.6.6). This insight requires further research.

We analyzed the role of professional development content in science teachers' innovative work behavior (Fig. 3.6.1) in the TIMSS 2015 data. We found that professional development content science curriculum and inquiry is statistically significantly associated with all the components of science teachers' innovative work behavior: generation, championing, implementation, and sharing of new ideas. We examined only one item, which is application of innovations in science education—problem-solving activity. We believe it is important to explore the associations between professional development content and other innovative activities: experimental, project-based, and other innovative activities. It is likely that the nature of innovative activities is related to different professional development components.

Rogers' diffusion theory (Rogers, 2003) highlighted the main stages in innovative processes. We found that the professional development content itself was important in the first (generating new ideas) and last (sharing new ideas) stages of the innovation emergence process: science curriculum, inquiry and critical thinking, and integrating information and communication technology (Fig. 3.6.1). The following professional development contents were also important in the second (championing new ideas) and third (implementation of new ideas) stages of the innovation emergence process: science curriculum, science pedagogy, improving students' critical thinking or inquiry skills, and integrating information and communication technology (ICT) into science.

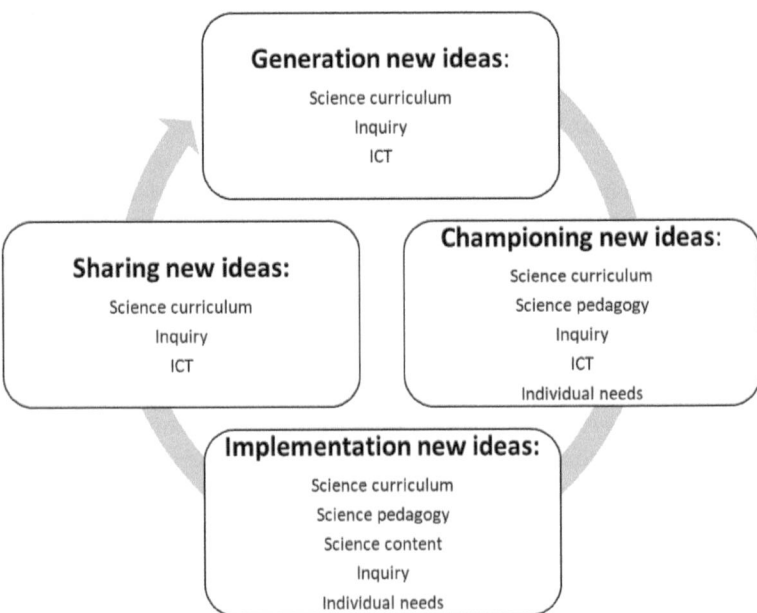

**Fig. 3.6.1:** The associations between components of science teachers innovative work behavior and professional development curriculum

We observed from TIMSS 2015 data that the professional development item "science assessment" had no role to play in any of the three countries chosen for the study (Singapore, Sweden, Lithuania). Therefore, new research is needed to verify the item science assessment's associations with components of innovative work behavior: generation, championing, implementation, and sharing of new ideas.

We explored the associations between science teachers 'gender and their innovative work behavior activities: generation, championing, implementation, and sharing of new ideas. However, in only one case (Lithuania) did we find a statistically significant association between championing of new ideas and the gender of the teacher. As mentioned earlier, this dependence may have been due to the gender-based variance in the proportion of teachers from Lithuania (Tab. 3.6.6). The Lithuanian female teachers who participated in the TIMSS 2015 study were seven times more in number than the male teachers who participated in the study. Further research on the associations between teachers' gender and innovative work behavior is needed.

### 3.6.5. Conclusions

The influence of professional development duration on science teachers' innovative work behavior is more the exception than the rule. In the case of only one country (Lithuania) and only one activity (championing), a statistically significant association between the two was established.

We found a statistically significant effect of science teachers' professional experience (PE) on all innovative activities: generation, championing, implementation (problem-solving activity), and sharing of new ideas.

The professional development item "science curriculum and inquiry" is statistically significantly associated with all components of the innovative work behavior of science teachers: generating, championing, implementing, and sharing of new ideas.

## 4. Discussion

This study aims to explore science teachers' innovative work behavior in the light of diffusion theory (Rogers, 2003). The Trends in International Mathematics and Science Study (TIMSS 2015) data on teachers from Lithuania, Sweden, and Singapore were analyzed through a linear modeling based on diffusion theory: (1) knowledge, (2) persuasion, (3) decision, (4) implementation, and (5) confirmation (Rogers, 2003). The TIMSS 2015 instrument for science teachers allowed to analyze science teachers' innovative work behavior abilities at the stages of idea generation (knowledge, persuasion, decision), implementation, and confirmation.

Hsiao et al. (2011) state that "teachers usually have good ideas, but they still need to discuss with their peers or supervisors, and they even attempt to convince each other. Once in a while, when the teachers' new idea is adopted, the realization of an innovative idea has begun. As a result, it may help teachers create higher innovative work behavior step by step" (Hsiao et al., 2011, p. 34). From our analysis of the data for Lithuania, we found that science teachers' new idea implementation abilities are better than their ability to generate new ideas.

Marth et al. (2018) analyzed the relationship between science motivation, technology interest, and innovative behavior of science teachers in a summer science school. Marth et al. (2018) talk of inquiry-based learning like innovative learning and believe that science teachers' innovative behavior is a powerful tool for promoting school students' motivation and interest in science and technology. Scholars observe that if "teachers are highly motivated and interested in science as well as in technology by furthermore using new methods like IBSE approaches in the classroom, a spill over to students seems likely" (Marth et al., 2018, p. 58). We analyzed how science teachers apply innovation in classroom by asking students to decide their own-problem-solving procedures. The results of our study revealed that 45.9% of Lithuanian science teachers' *often* or *very often* ask students to decide their own problem-solving procedures. It was surprising that 93.01% of science teachers in *every lesson*, in *almost every lesson*, or in *about half the lessons* encourage students to express their new ideas in classroom. But we didn't analyze how science teachers' innovative work behavior influences students' motivation for learning science, their interest in technology, and their innovative work behavior as students.

We have done a statistical analysis of science teachers' innovative work behavior abilities using the path analysis procedure with AMOS 17. Our empirical model of science teachers' innovative work behavior abilities using path analysis procedure with AMOS addressed the following: generating new ideas in teaching science → development of new ideas → applying new ideas in education → promotion of new ideas in education → sharing new ideas about teaching science (Tab. 3.1.6). We also analyzed one indirect way: Development of new ideas indirectly affects science teachers' ability to promote new ideas in education (Fig. 3.1.1). The paths in our empirical model (Fig. 3.1.1) correspond to the channels in Rogers (2003) model. Path analysis with AMOS confirmed the diffusion theory. All direct paths were significant in the final model (Fig. 3.1.1). The ability of science teachers' to generate new ideas in teaching science had a statistically significant effect on their ability to develop new ideas. The ability to develop new teaching ideas had a statistically significant influence on the ability to apply and promote new ideas in education. The ability to promote new ideas in education had a statistically significant effect on the ability to share new ideas about teaching science. As mentioned earlier, the most significant effects observed in the model include the following: the effect of promotion of new ideas in science education encouraging students to express their ideas in classroom on the sharing of new teaching ideas ($\beta = 0.992$); the effect of applying new ideas in problem-solving process on science teachers' ability to promote new ideas in education ($\beta = 0.782$); the effect of generating new ideas about teaching on science teachers' ability to develop new ideas ($\beta = 0.703$.).

In the present study, we analyzed the influence of science teachers' demographic factors (age and gender) and the education factor on their innovative work behavior (trying out new ideas and sharing new ideas) in education. For the purpose of this study, we chose a quantitative analysis of data based on ordinal logistic regression (OLR). Thurlings et al. (2015) conducted extensive literature analysis on the role of demographic factors in teachers' innovative work behavior and found that "an overview of the demographic factors, of which all except one were found in quantitative research" (Thurlings et al., 2015, p. 14).

Scholars not very often apply regression analysis to investigate teachers' innovative work behavior (Thurlings et al., 2015). We applied an OLR analysis in order to determine the influence of science teachers' demographic, professional, and educational factors on their innovative work behavior, specifically their ability to try out and share new ideas in science education. According to Thurlings et al. (2015), a wide range of demographic factors have been studied: gender, age, income, level of education, years of education, and teaching experience. "Only the variables, income, years of education, and having other functions had a significant positive effect on innovative work behavior" (Thurlings et al., 2015, p. 14).

## Discussion

The most important contribution of our study was to validate a mathematical model that incorporates demographic and educational characteristics as a predictors of science teachers' innovative work behavior, specifically the ability to try out new ideas and share those new ideas. We found that a short-cycle tertiary level (EDU-4; Tab. 3.2.5) has a negative influence on the science teachers' ability to generate new ideas. Our results are in line with that of Yang and Huang (2008): "higher levels of computer training, computer literacy, well-supported school environment [...] result in higher task intensity, impact concerns and more technology-mediated teaching behaviour in the classroom" (Yang & Huang, 2008). It should be noted that these researchers (Yang & Huang, 2008) used only one aspect—the application of information and communication technology (ICT)—to analyze innovations in education. However, it was found that education level did not have a statistically significant influence on science teachers' ability to share new ideas (Tab. 3.2.6).

The study has two main limitations. First limitation is that the TONI-1 model is based on national-level data and only the data for science teachers from Lithuania were analyzed. Future studies could replicate this research with international-level data and use mathematical models based on data for other countries that participated in the TIMSS 2015.

The second limitation of this study is that the TONI-1 model did not have high pseudo $R^2$ values (e.g., Nagelkerke = 21%). It indicates that the predictors EDU, AGE, GEN explain only a relatively small proportion of the variation between science teachers' ability to generate new ideas in education. This turned out to be just what we expected because there were numerous characteristics that we found had an impact on science teachers' innovative work behavior. These will be addressed in other sections of this monograph.

The present study examined the influence of organizational/affective commitment on science teachers' innovative work abilities. We adopted the exploratory factor analysis (EFA) and confirmatory factor analysis (CFA) methods and path analysis to test the structural model. The results indicated that science teachers' organizational/affective commitment had a significant influence on their innovative work generation of new ideas in education, development of new ideas in education, applying new ideas in education, promotion of new ideas in education, and modification and sharing of new ideas.

Applying new ideas in education is related to problem-solving (Kim et al., 2018). We analyzed science teachers' ability to ask students to decide their own problem-solving procedures. The structural equation modeling (SEM) analysis of the TIMSS 2015 data on science teachers revealed a statistically significant influence of organization/affective commitment on science teachers' ability to ask students to decide their own problem-solving procedures. Chang (2018) analyzed the effects of teachers' knowledge innovation on students' innovative work behavior by using a hierarchical linear modeling. It was found that teachers' knowledge innovation had a positive moderation effect on students' innovation behaviors (Chang, 2018).

Applying new ideas in education works at two levels: the teacher level and the student level (Chang, 2018). We analyzed the teacher level on the basis of their organization/effective commitment and the student level based on teachers asking students to decide their own-problem solving procedures. Chang (2018) analyzed the teacher level using teachers' knowledge innovation and the student level based on students' innovation behavior. Results of our research about applying of new ideas in science education confirm the importance of the two-level model: teacher level and student level.

We analyzed science teachers' innovative work behavior based on their relations and interactions with other people, such as working together to try out new ideas and visiting another classroom to learn more about teaching. According to Thurlings et al. (2015), these categories belong to the organizational factor classifications (interactions support, guidance, and feedback collegiality). We found that organizational/affective commitment had a major influence on science teachers' ability to work together ($\beta = 0.375$, $p < 0.05$) and their visiting another classroom to learn more about teaching ($\beta = 0.414$, $p < 0.05$). Our results correspond to the results reported by other scholars. Messmann and Mulder (2011) highlights the social aspect of innovative work behavior by stating that the development of innovations is a complex, iterative, and primarily social process. Widmann and Mulder (2018) state that interaction, especially reflexivity, is related positively to innovative work behavior.

We found that organizational/affective commitment influences science teachers' innovative work behavior. It can be assumed that teachers' innovative work behavior leads to innovation within the organization. Ming and Ying (2010) analyzed how affective commitment influences technological innovation and administrative innovation. They applied the SEM analyze whether and how the affective commitment influences organizational innovation. Ming and Ying (2010) found that affective commitment had a direct, significant, and positive effect on both technological innovation and administrative innovation. A lot of research remains to be done to study the simultaneous influence of the three organizational commitments on science teachers' innovative work behavior abilities.

In the present study, we analyzed the influence of science teachers' self-confidence in teaching science on their innovative work behavior abilities. We investigated teachers' innovative work behavior by carrying out a secondary analysis of the TIMSS 2015 data. We used the EFA for investigating the variable relationships of complex concepts, including, science teachers' innovative work behavior. The EFA revealed two factors: idea generating and sharing and idea applying. The SEM demonstrated that path coefficients are lower in the variables group of idea generating and sharing (Tab. 3.4.9). It means science teachers' self-confidence in teaching science had less influence on their ability to generate and share new ideas and more influence on their ability to apply those new ideas. Serdyukov (2017) posits that "innovation requires three major steps: an idea, its implementation, and the outcome that results from the execution of the idea and produces a change" (p. 8). The SEM results showed that science teachers' self-confidence in teaching science had more influence on their innovative work behavior at the second stage of the innovation process—the idea implementation stage.

We analyzed the influence of science teachers' self-confidence in teaching science on their innovative work behavior at the individual level. Scholars state that the field of innovation covers activities ranging from the organizational level down to the individual level (Axtell et al., 2000). At the individual level, innovation occurs in terms of the implementation of small-scale ideas related to bringing in improvements in daily work processes and activities (Axtell et al., 2000). In the present study, teachers' innovative work behavior is studied at the individual level, as teachers primarily contribute to small-scale innovations in their work roles and initiate the process of innovation in teaching. It would make sense to repeat our study at the next level, which is the organization.

Other limitations of our research include the difficulty in measuring innovative work behavior. "Both scientists and practitioners emphasize the importance of innovative work behavior (IWB) of individual employees for organizational success, but the measurement of IWB is still at an evolutionary stage" (De Jong & Hartog, 2010, p.23). We did not use a special questionnaire to investigate the innovative work behavior of science teachers. We conducted a secondary analysis of science teachers' innovative work behavior using the TIMSS 2015 questionnaire.

The present research considered the processes activated in science teachers' innovative work behavior. We have shown how organization leadership support (OLS) is useful for promoting science teachers' innovative work behavior.

We investigated the latent construct *organization leadership support* on the basis of TIMSS 2015 questionnaire. We used an EFA to investigate the variables of OLS. The EFA revealed the group of questions related to OLS: teachers' understanding of school's curricular goals, collaboration between school leadership and teachers to plan instruction, amount of instructional support provided to teachers by school leadership, and school leadership's support for teachers' professional development. We carried out a confirmatory factor analysis (CFA) to confirm our EFA results about organization leadership support. We found the highest loading for the variable collaboration between school leadership and teachers to plan instruction. This variable expresses relational leadership. Leadership's scholars state. that "leadership is not based only at the traits of the leader but it is the social process that occurs between followers and the leader, i.e. leadership is a relational process" (Akram et al., 2016). Our research confirms the role of relational leadership in the latent variable OLS ($\beta = 0.761$, $p < 0.001$).

The role of leadership in innovation is explored not only in education. Akram et al. (2016) investigated the effect of relational leadership on employee innovative work behavior in the information technology industry of China. Scholars observed that relational leadership helps employees to demonstrate innovative work behavior at each stage of the idea generation process. Results of our research in education area confirm the main conclusion about the influence of leadership on innovation process in technology industry in different cultures: "In order to improve the employee innovative work behavior, organizations need to promote relational leadership among their leader" (Akram et al., 2016).

The moderator role of non-educational organization support for innovative work behaviors was analyzed by Mokhber et al. (2018). They reported a positive relationship between transformational leadership and organizational innovation according to the main components of innovative work behavior: idea generating, risk-taking, and decision-making. Scholars state that transformational leaders might not only promote innovative activity within the organization but also ensure the market success of the innovations (Mokhber et al., 2018,). In our study, we have addressed the role of only transformational leaders for innovation in the organization but did not consider any other type of leadership. However, our study confirmed the following trends: OLS influences the innovative work behavior of science teachers. The extent and scope of our research results are broader and do not cover only one type of leadership. We believe further research is needed to study in detail the different types of leadership: transversal, shared, and so on.

## Discussion about the role of predictors: DPD, CPD, PE, and GEN

To address three research hypotheses, we investigated in what way the four predictors DPD (duration of professional development, in hours), CPD (content of professional development), PE (professional experience—number of years the teacher has been teaching—and its influence on the probability of science teachers implementing innovations in science education collaboratively), and GEN (teacher's gender).

Researchers posit that sustained duration is an important feature of professional development, and this, in turn, leads to the conclusion that there is no alternative to sustained duration in professional development (Bautista, Wong, & Gopinathan, 2015; Bilgin & Balbag, 2018; Darling-Hammond et al., 2009; Desimone, 2009; Knapp, 2003; Weiss & Pasley, 2006; Zulu & Bertram, 2019). However, other researchers have voiced frustration that the research has not yet identified a clear threshold how long the duration should be on average for professional development models to be effective (Darling-Hammond et al., 2009, p. 13). The testing of different math models based on OLR can provide an opportunity to determine a clear threshold value for the duration of professional development; that is, how long at a bare minimum should professional development courses last to be effective?

Using the TIMSS 2015 data for Singapore, Lithuania, and Sweden, we performed an extensive OLR analysis of the role and influence of professional development and professional and demographic factors in the innovative work behavior of science teachers. Results confirmed that the importance of professional development to science content and science pedagogy in constructivist approaches is unquestionable (Aldahmash, Alshamrani, Alshaya & Alsarrani, 2019; Bantwini, 2019; Chai, 2019).

Sustained duration is an important feature of professional development; however, empirical evidence regarding positive impacts is controversial. "Though research has not yet identified a clear threshold for the duration of effective PD models, it does indicate that meaningful professional learning that translates to changes in practice cannot be accomplished" (Darling-Hammond et al., 2009, p. 13).

Some research indicates that professional development activities need sufficient duration and that meaningful professional learning that translates to changes in practice cannot be accomplished in short, one-off workshops (Darling-Hammond et al., 2009; Desimone, 2009; Knapp, 2003; Weiss & Pasley, 2006). From the TIMSS 2015 data for Lithuania, we found that DPD had an influence on science teachers' activity of championing new ideas. We confirmed the information that science teachers who spent 16 or more hours of professional development in two years were more likely to champion new ideas in science education. In the case of only one country (Lithuania) and only one activity (championing), a statistically significant association was established between duration of professional development and the activity of championing new ideas.

Scholars analyzed the association between teachers' professional experience and collaborative innovative activity (Brekelmans et al., 2005; Hargreaves, 2005; Ertesvåg, 2014). Teachers' professional experience, teaching skills, and disposition influence the educational practice (Kaya & Elster, 2019). In general, teachers gain professional experience in a process that begins with pre-service training and continues with in-service training (Kaya & Gödek, 2016). Brekelmans, Wubbels, and van Tartwijk (2005) discussing about role of teachers experience in education revealed two factors: time for acquiring the skills for tackling challenges and enthusiasm about bringing in changes. Less experienced teachers may take a few more years before they acquire the abilities to manage challenges and develop and implement innovations, but it is equally true that in general only those teachers early in their career demonstrate more enthusiasm in facing and overcoming challenges compared to their senior counterparts (Brekelmans, Wubbels, & van Tartwijk, 2005).

Hargreaves (2005) argues that experienced teachers are more relaxed and feel more comfortable in their professional activity. In fact, more experienced teachers are likely to become resistant toward change at the end of their career (Hargreaves, 2005).

We found that science teachers' professional experience had a statistically significant effect on all innovative activities: generating, championing, implementing (problem-solving), and sharing new ideas. It should be noted that there was no statistically significant association between the components of teachers' professional experience and innovative work behavior (trying out, championing, implementing, and sharing new ideas) in the data for Singapore. This result may have been due to the significantly lower level of professional experience of Singapore teachers over the years (Tab. 3.6.6). This insight requires further research.

Using the TIMSS 2015 data, we analyzed the role of professional development content in science teachers' innovative work behavior (Fig. 3.6.1). We found that professional development content "science curriculum and inquiry" is statistically significantly associated with all of the components of science teachers' innovative work behavior: generation, championing, implementing, and sharing new ideas. We examined only one case of application of innovations in science education—problem-solving activities. We believe it is important to explore the associations between PD content and other innovative activities: experimental, project-based, and other innovative activities. It is likely that the nature of innovative activities is related to different PD contents.

Rogers' diffusion theory (Rogers, 2003) highlighted the main steps in the innovative processes. We found that the content of PD itself was important in the first (generating new ideas) and last (sharing new ideas) stages of the innovation emergence process: science curriculum, inquiry and critical thinking, and integrating information and communication technology (Fig. 3.6.1). The same PD content is important in the second (championing new ideas) and third (implementation of new ideas) stages of the innovation emergence process: science curriculum, science pedagogy, improving students' critical thinking or inquiry skills, and integrating information and communication technology (ICT) into science.

# 5. Conclusions

1. Educational innovation plays an important role in society. The implementation of innovations in practice depends on the person's innovative work behavior. The permanent relationship between innovation and science education has been observed for a long time. Innovation has become a permanent feature of science education, not only in curriculum content but also in the associated teaching methods and materials.
Rogers' theory of diffusion of innovations is widely used in investigating the spread of innovations in different contexts, including in education. Very often it is used in analyzing technological innovations in education. The theory is also used in investigating teachers' innovative practices and innovative work behavior in secondary education and at other levels of the education system.
Innovative work behavior is understood as a multidimensional construct in which individuals generate novel ideas or schemes in their work, solve practical problems at work, and achieve positive effects, including the generation, development, and implementation of new ideas. Innovative work behavior depends on personal cognitive and environmental factors and actors.
2. Science teachers of Lithuania demonstrate the ability to apply innovations in the practice. A comparative analysis of the highest ranks percentages sum confirmed that science teachers' idea generating abilities (from the first stage to the third stage, according to diffusion theory) are weaker than the abilities to implement new ideas and confirm those new ideas. A statistically significant difference was detected between all of the science teachers' innovative work behavior abilities.
A statistically significant correlation was found between science teachers' ability to try out new idea and their ability to look for new ideas by visiting other classes and interacting with other teachers; between their ability to look for new ideas and their ability to share new ideas; and between their ability to try out new ideas and their ability to share new ideas.
The path analysis TIMSS 2015 data of science teachers' innovative work behavior components confirmed the Diffusion theory. The influence of the promotion of new ideas in science education by encouraging students to express their ideas in classroom on teachers sharing new teaching ideas was most significant. Applying new ideas in problem-solving processes has a big and statistically significant effect on science teachers' abilities to promote new

ideas in education. The generation of new ideas about teaching has a statistically significant effect on science teachers' ability to develop new ideas.
3. The following conclusions can be drawn from the mathematical model TONI-1 of this research which was created based on TIMSS 2015 data of science teachers from Lithuania: (1) science teachers trying out new ideas collaboratively depends on the level of education and the gender of the teacher; (2) the short-cycle tertiary level (EDU-4) has a negative influence on the ability of science teachers' to try out new ideas collaboratively in education; (3) the gender of science teachers has a positive influence on the ability of science teachers' to try out and share new ideas in education. The female teachers are more likely to try out and share new ideas than their male counterparts.
4. In this book, we strove to provide the most reliable portrait possible of the influence of science teachers' organizational/affective commitment on their innovative work behavior. Using a comparative factor analysis (CFA) and structural equation modeling (SEM) of the TIMSS 2015 data of Lithuania, we found that organizational/affective commitment has a statistically significant influence on science teachers' innovative work behavior.

The CFA analysis revealed that organizational/affective commitment is described by a set of TIMSS 2015 questionnaire variables: I am content with my profession as a teacher; I am satisfied with being a teacher at this school; I find my work full of meaning and purpose; I am enthusiastic about my job; My work inspires me; I am proud of the work I do; I am going to continue teaching for as long as I can.

The SEM disclosed the different levels of influence of organizational/affective commitment on science teachers' innovative work behavior abilities. Organizational/affective commitment had the biggest influence on science teachers' ability to share innovative ideas ($\beta = 0.419$, $p < 0.05$) and the lowest influence on applying new ideas in education ($\beta = 0.331$, $p < 0.05$).
5. Science teachers implement learning strategies through various activities. A statistical analysis of various activities (inspiring students to learn science, explaining science concepts or principles by doing science experiments, providing challenging tasks for the highest achieving students, engaging students' interest, helping students appreciate the value of learning science, assessing students' comprehension of science, improving the understanding of struggling students, making science relevant to students, developing students' higher-order thinking [HOT] skills, and teaching science using inquiry methods) revealed that science teachers' self-confidence in teaching science using inquiry methods had the most influence on science teachers' confidence in teaching science. Science teachers' activity of helping students

appreciate the value of learning science had the least influence on science teachers' confidence in teaching science.

The present study found that science teachers' self-confidence in teaching science has a statistically significant and positive influence on the innovative work behavior of science teachers. The EFA revealed two groups of science teachers' innovative work behavior abilities: idea generating and sharing, and idea applying.

The SEM results confirmed the influence of science teachers' self-confidence in teaching science on their innovative work behavior abilities. Self-confidence of science teachers in teaching science had more influence on the idea applying ability and less influence on idea generating and sharing abilities.

Analysis of path coefficients for the second factor (idea applying) confirmed that science teachers' innovative ability to encourage students to express their ideas in class and their innovative ability to ask students to decide their own problem solving procedures are directly and positively affected by science teachers' self-confidence in teaching science.

6. Findings from the empirical analysis carried out in the present study show a statistically significant influence of organization leadership support on science teachers' innovative work behavior abilities.

It was found that organization leadership support had more influence on science teachers' ability to generate new ideas and the ability to share those new ideas and less influence on the ability to applying new ideas in practice. Organization leadership support had a statistically significant effects on science teachers' abilities to work together in trying out new ideas and to share what they have learned.

7. The influence of professional development duration on science teachers' innovative work behavior is more the exception than the rule. In the case of only one country (Lithuania) and only one activity (championing), a statistically significant association was established between professional development duration and innovative work behavior. Science teachers' professional experience (PE) had a statistically significant effect on all innovative activities: generation, championing, implementation (problem-solving activity), and sharing of new ideas. The professional development content "science curriculum and inquiry" is statistically significantly associated with all the components of the science teacher's innovative work behavior: generating, championing, implementing, and sharing new ideas.

# Approval

The monograph SCIENCE TEACHERS' INNOVATIVE WORK BEHAVIOR: FACTORS AND ACTORS was approved for publishing by the Commission of the Institute of Educational Research (Decision No. ET2-4) and by The Council of The Education Academy (Decision No. 61).

# List of figures

| | | |
|---|---|---|
| **Fig. 1.3.1:** | Modeling individual's behavior: social learning theory | 26 |
| **Fig. 1.3.2:** | Innovative work behavior: social learning theory and diffusion theory | 27 |
| **Fig. 1.3.3:** | Innovative work behavior: social cognitive theory | 28 |
| **Fig. 1.3.4:** | Motivation for innovations: Expectancy-value theory | 30 |
| **Fig. 3.1.1:** | The structure of science teachers' innovative work behavior: unstandardized coefficients | 47 |
| **Fig. 3.3.1:** | The structural equation model: science teachers' organizational/affective commitment (OAC) and their innovative work behavior | 73 |
| **Fig. 3.4.1:** | Conceptual framework of science teachers' self-confidence in teaching science and innovation behavior | 81 |
| **Fig. 3.4.2:** | SEM results of science teachers' innovative work behavior abilities and their confidence in teaching science: unstandardized data | 87 |
| **Fig. 3.5.1:** | The model of the influence of organization leadership support on science teachers' innovative work behavior | 95 |
| **Fig. 3.5.2:** | Conceptual framework of school organization leadership support | 97 |
| **Fig. 3.5.3:** | SEM framework: conceptual framework of science teachers' innovative work behavior abilities and organization leadership support | 98 |
| **Fig. 3.5.4:** | SEM results of science teachers' innovative work behavior abilities and organization leadership support: organizational leadership support | 103 |
| **Fig. 3.6.1:** | The associations between components of science teachers' innovative work behavior and professional development curriculum | 150 |

# List of tables

| | | |
|---|---|---|
| Tab. 3.1.1: | TIMSS 2015 questions about science teachers' innovative behavior | 39 |
| Tab. 3.1.2: | Results of Kolmogorov–Smirnov test and values of asymmetry | 40 |
| Tab. 3.1.3: | Values of data asymmetry | 40 |
| Tab. 3.1.4: | Innovative work behavior (generation-development-sharing) of science teachers: percentage frequency | 41 |
| Tab. 3.1.5: | Innovative work behavior of science teachers in the classroom: percentage frequency | 42 |
| Tab. 3.1.6: | Science teachers' innovative work behavior abilities: sum of percentages of two highest ranks (very often, often/every or almost every lesson, about half the lessons) | 43 |
| Tab. 3.1.7: | Spearman correlation between science teachers' innovative work behavior components | 44 |
| Tab. 3.1.8: | Fitness of items of science teachers' innovative behavior abilities | 45 |
| Tab. 3.1.9: | The structure and system of science teachers' innovative behavior abilities: direct and indirect path coefficients and statistical significance | 47 |
| Tab. 3.2.1: | The hypothesis testing about science teachers' innovative work behavior: TONI and SHNI models | 54 |
| Tab. 3.2.2: | Model fitting information derived from the ordinal regression analysis: to explain science teachers' ability to try out new ideas and share new ideas in science education | 56 |
| Tab. 3.2.3: | Goodness-of-fit test: to explain science teachers' ability to try out new ideas and share new ideas in science education | 57 |
| Tab. 3.2.4: | Pseudo $R^2$ Nagelkerke: to explain science teachers' ability to try out new ideas and share new ideas in science education | 57 |
| Tab. 3.2.5: | Parameter estimates to explain science teachers' ability to try out new ideas in science education (model TONI-1) | 58 |
| Tab. 3.2.6: | Test of parallel lines: to explain science teachers' ability to try out new ideas and share new ideas in science education | 59 |
| Tab. 3.2.7: | Cross-tabulation of variables: to explain science teachers' ability to try out new ideas in science education (model TONI-1) | 60 |
| Tab. 3.2.8: | Parameter estimates: to explain science teachers' ability to share new ideas in science education (model SHNI-1) | 61 |

# List of tables

| | | |
|---|---|---|
| Tab. 3.2.9: | Cross-tabulation of variables: to explain science teachers' ability to share new ideas in science education (model SHNI-1) | 62 |
| Tab. 3.2.10: | The results of hypothesis testing | 63 |
| Tab. 3.3.1: | Science teachers' organizational/affective commitment (based on TIMSS 2015 data) | 67 |
| Tab. 3.3.2: | The skewness and kurtosis values of the items of science teachers' organizational/affective commitment | 67 |
| Tab. 3.3.3: | Component matrix of organizational/emotional commitment scale. Correlations among the items of the affective commitment scale | 68 |
| Tab. 3.3.4: | Intercorrelations between the items of affective commitment | 69 |
| Tab. 3.3.5: | The fitness of items of the organizational/affective commitment factor | 69 |
| Tab. 3.3.6: | Standardized and unstandardized coefficients of the organizational/affective commitment factor | 70 |
| Tab. 3.3.7: | Science teachers' innovative work behavior abilities and the stages of the diffusion theory | 71 |
| Tab. 3.3.8: | Results of SEM: influence of organizational/affective commitment on science teachers' innovation behavior | 73 |
| Tab. 3.4.1: | Science teachers' confidence in teaching science: strategies and outcomes | 79 |
| Tab. 3.4.2: | Normality of the data on science teachers' confidence in teaching science: Kolmogorov–Smirnov test | 80 |
| Tab. 3.4.3: | Normality of the data on science teachers' confidence in teaching science: asymmetry coefficients test | 80 |
| Tab. 3.4.4: | Anti-image correlation matrix of questions about science teachers' self-confidence in teaching science | 83 |
| Tab. 3.4.5: | Eigenvalues percentage of variance and cumulative percentages for the factor science teachers' self-confidence in teaching science | 83 |
| Tab. 3.4.6: | Observed variables and loadings of the factor science teachers' self-confidence in teaching science | 84 |
| Tab. 3.4.7: | The fitness of items of the latent factor science teachers' self-confidence in teaching science | 85 |
| Tab. 3.4.8: | Standardized and unstandardized coefficients of the latent variable science teachers' self-confidence in teaching science: the latent construct "science teachers' confidence in teaching science" (CTS) | 85 |

# List of tables

| | | |
|---|---|---|
| Tab. 3.4.9: | The influence of science teachers' self-confidence in teaching science on science teachers' innovative work behavior abilities: path coefficients and statistical significance | 88 |
| Tab. 3.5.1: | The influence of organization leadership support on science teachers' innovative work behavior variables (based on TIMSS 2015 data) | 95 |
| Tab. 3.5.2a: | The skewness and kurtosis values of organization leadership support variables | 96 |
| Tab. 3.5.2b: | The skewness and kurtosis values of organization leadership support variables after removing extreme values | 96 |
| Tab. 3.5.3: | Anti-image correlation matrix of questions about organization leadership support | 100 |
| Tab. 3.5.4: | Eigenvalues percentage of variance and cumulative percentages for factor of the organization leadership support | 100 |
| Tab. 3.5.5: | Observed variables and loadings of the factor organization leadership support | 100 |
| Tab. 3.5.6: | Anti-image correlation matrix of questions about science teachers' innovative work behavior abilities | 101 |
| Tab. 3.5.7: | The results of the principal component analysis (PCA) of science teachers' innovative behavior | 102 |
| Tab. 3.5.8: | Factor analysis table of science teachers' innovative work behavior | 102 |
| Tab. 3.5.9: | The influence of organization leadership support on science teachers' innovative work behavior abilities: path coefficients and statistical significance | 104 |
| Tab. 3.6.1: | Main features of initial teacher education in the three countries chosen for the study: Lithuania, Sweden, and Singapore | 113 |
| Tab. 3.6.2: | TALIS 2018 indicators for initial teacher education and novice period in the countries chosen for the study: Lithuania, Sweden, and Singapore | 114 |
| Tab. 3.6.3: | Main features of continuous teacher education in the countries chosen for the study: Lithuania, Sweden, and Singapore | 115 |
| Tab. 3.6.4: | TALIS 2018 indicators for continuous teacher education in the countries chosen for the study: Lithuania, Sweden, and Singapore | 115 |

| | | |
|---|---|---|
| Tab. 3.6.5: | The percentage of science teachers' participation in professional development: the aspect of content | 117 |
| Tab. 3.6.6: | Descriptive statistics of science teachers: sample, the number of years the teacher has been teaching, and the gender of the teacher | 117 |
| Tab. 3.6.7: | The time science teachers spent in professional development in the last two years: Singapore, Lithuania, and Sweden | 118 |
| Tab. 3.6.8: | Contents of science teachers' professional development | 119 |
| Tab. 3.6.9: | Model (1) fitting information from an ordinal regression analysis: to explain science teachers trying out new ideas in science education | 120 |
| Tab. 3.6.10: | Parameter estimates to explain science teachers' activity of trying out new ideas in science education: based on the data for Lithuania | 122 |
| Tab. 3.6.11: | Parameter estimates to explain science teachers' activity of trying out new ideas in science education: based on the data for Sweden | 123 |
| Tab. 3.6.12: | Parameter estimates for the TONI model: Singapore | 124 |
| Tab. 3.6.13: | Test of parallel lines: to explain science teachers' activity of trying out new ideas in science education: based on the data for Lithuania, Sweden, and Singapore | 125 |
| Tab. 3.6.14: | Science teachers' professional development course contents that had a statistically significant effect: based on the data for Lithuania, Sweden, and Singapore | 126 |
| Tab. 3.6.15: | Model (2) fitting information from an ordinal regression analysis: to explain science teachers' activity of championing new ideas in science education | 127 |
| Tab. 3.6.16: | Test of parallel lines: to explain science teachers' activity of championing new ideas in science education | 127 |
| Tab. 3.6.17: | Parameter estimates to explain science teachers' activity of championing new ideas in science education: Lithuania | 128 |
| Tab. 3.6.18: | Cross-tabulation of variables based on the data for Lithuania: championing of new ideas and the number of years the teacher has been teaching | 130 |
| Tab. 3.6.19: | Parameter estimates to explain science teachers' activity of championing new ideas in education: Sweden | 131 |
| Tab. 3.6.20: | Cross-tabulation of variables based on the data for Sweden: championing of new ideas and the number of years the teacher has been teaching | 132 |

| | | |
|---|---|---|
| Tab. 3.6.21: | Parameter estimates to explain science teachers' activity of championing new ideas in education: Singapore | 133 |
| Tab. 3.6.22: | The content of science teachers' professional development that had a statistically significant effect on their championing new ideas: based on the data for Lithuania, Sweden, and Singapore | 134 |
| Tab. 3.6.23: | Model (PPS) fitting information from an ordinal regression analysis | 136 |
| Tab. 3.6.24: | Parameter estimates to explain science teachers promoting problem-solving activity in science education: Lithuania | 137 |
| Tab. 3.6.25: | Cross-tabulation of the variables "promoting problem-solving" and "the number of years the teacher has been teaching" (professional experience [PE]): Lithuania | 138 |
| Tab. 3.6.26: | Parameter estimates to explain science teachers promoting problem-solving activity in science education: Sweden | 139 |
| Tab. 3.6.27: | Parameter estimates to explain science teachers promoting problem-solving activity in science education: Singapore | 140 |
| Tab. 3.6.28: | Test of parallel lines to explain the PPS model: Lithuania, Sweden, and Singapore | 141 |
| Tab. 3.6.29: | The content of science teachers' professional development that had a statistically significant influence on science teacher's promoting problem-solving in education: Lithuania, Sweden, and Singapore | 142 |
| Tab. 3.6.30: | The SHARE model fitting information from an ordinal regression analysis: Lithuania, Sweden, and Singapore | 143 |
| Tab. 3.6.31: | Parameter estimates for the SHARE model: Lithuania | 144 |
| Tab. 3.6.32: | Cross-tabulation of variables on the data for Sweden: sharing new ideas and the number of years the teacher has been teaching | 145 |
| Tab. 3.6.33: | Parameter estimates for the SHARE model: Sweden | 146 |
| Tab. 3.6.34: | Parameter estimates for the SHARE model: Singapore | 147 |
| Tab. 3.6.35: | Test of parallel lines for the SHARE model: Lithuania, Sweden, and Singapore | 148 |
| Tab. 3.6.36: | The content of science teachers' professional development that had a statistically significant effect on science teachers sharing new ideas in education: Lithuania, Sweden, and Singapore | 148 |

# Bibliography

Abdullah, A. (2011). Evaluation of Allen and Meyer's organizational commitment scale: A cross-cultural application in Pakistan. *Journal of Education and Vocational Research 1*(3), 80–86.

Adams Becker, S., Brown, M., Dahlstrom, E., Davis, A., DePaul, K., Diaz, V., & Pomerantz, J. (2018). *NMC horizon report: 2018 higher education edition*. Louisville, CO: EDUCAUSE, 2018. Retrieved from from https://library.educause.edu/~/media/files/library/2018/8/2018horizonreport.pdf

Agarwal, P. K., & Roediger, H. L. (2018). Lessons for learning: How cognitive psychology informs classroom practice. *Phi Delta Kappan, 100*(4), 8–12.

Agarwal, P. K., Bain, P. M., & Chamberlain, R. W. (2012). The value of applied research: Retrieval practice improves classroom learning and recommendations from a teacher, a principal, and a scientist. *Educational Psychology Review, 24*, 437–448.

Akram, T., Lei, S., & Haider, M. J. (2016). The impact of relational leadership on employee innovative work behavior in IT industry of china. *The Arab Economic and Business Journal, 11*(2), 153–161. doi:10.1016/j.aebj.2016.06.001.

Aksela, M. (2019). Towards student-centred solutions and pedagogical innovations in science education through co-design approach within design-based research. *LUMAT: International Journal on Math, Science and Technology Education, 7*(3), 113–139.

Aldahmash, A. H., Alshamrani, S. M., Alshaya, F. S., & Alsarrani, N. A. (2019). Research trends in in-service science teacher professional development from 2012 to 2016. *International Journal of Instruction, 12*(2), 163–178. https://doi.org/10.29333/iji.2019.12211a

Amabile, T. M., Conti, R., Coon, H., et al. (1996). Assessing the work environment for creativity. *Academy of Management Journal, 39*, 1154–1184. http://dx.doi.org/10.2307/256995.

Appleton, K. (1992). Discipline knowledge and confidence to teach science: Self-perceptions of primary teacher education students. *Research in Science Education, 22*, 11–19.

Arici, F., Yildirim, P., Caliklar, Ş., & Yilmaz, R. M. (2019). Research trends in the use of augmented reality in science education: Content and bibliometric mapping analysis. *Computers & Education, 142*, 103647. https://doi.org/10.1016/j.compedu.2019.103647.

Axtell, C. M., Holman, D. J., Unsworth, K. L., Wall, T. D., Waterson, P. E., & Harrington, E. (2000). Shopfloor innovation: Facilitating the suggestion and implementation of ideas. *Journal of Occupational and Organizational Psychology, 73*, 265–285. doi:10.1348/096317900167029.

Aziah, I., & Al Amin, M. (2018). The impact of transformational leadership and commitment on teachers' innovative behaviour. In *Advances in Social Science, Education and Humanities Research (ASSEHR), volume 304.* 4th ASEAN Conference on Psychology, Counselling, and Humanities (ACPCH 2018), pp. 426–430.

Bada, S., & Prasadh, R. S. (2019). Professional Development of a Teacher for an Effective Teaching-Learning in School Education: A View. *Journal on Educational Psychology, 13*(1), 7–13.

Bandura, A. (1977a). Self-efficacy: Toward a unifying theory of behavioral change. *Psychological Review, 84*, 191–215. http://dx.doi.org/10.1037/0033-295X.84.2.191.

Bandura, A. (1977b). *Social learning theory.* Englewood Cliffs, NJ: Prentice Hall.

Bandura, A. (1994). Self-efficacy. In V. S. Ramachandran (Ed.), *Encyclopedia of human behavior* (Vol. 4, pp. 71–81). New York: Academic Press. (Reprinted in H. Friedman [Ed.], Encyclopedia of mental health. San Diego: Academic Press, 1998).

Bandura, A. (1997). *Self-efficacy: The exercise of control.* New York, NY: Y. H. Freeman.

Bandura, A. (2001). Social cognitive theory: An agentic perspective. *Annual Review of Psychology, 52*, 1–26.

Bandura, A. (2012). On the functional properties of perceived self-efficacy revisited. *Journal of Management, 38*, 9–44.

Bandura, A., & National Institute of Mental Health. (1986). *Social foundations of thought and action: A social cognitive theory.* Prentice-Hall, Inc.

Bantwini, B. D. (2019). Developing a culture of collaboration and learning among natural science teachers as a continuous professional development approach in a province in South Africa. *Teacher Development, 23*(2), 213–232. https://doi-org.ezproxy.vdu.lt:2443/10.1080/13664530.2018.1533491

Bardone, E., Burget, M., Saage, K., & Taaler, M. (2017). Making sense of responsible research and innovation in science education through inquiry-based learning. Examples from the field. *Science Education International, 28*(4), 293–304.

Barrow, E., Golding, J., Redmond, B., & Grima, G. (2018). "I get better and better all the time": Impact of resources on pupil and teacher confidence. In J. Golding, N. Bretscher, C. Crisan, E. Geraniou, J. Hodgen, & C. Morgan (Eds.), *Proceedings of the 9th British Congress of Mathematics Education: BCME9* (pp. pp. 8–15). BSRLM: University.

Bautista, A., Wong, J., & Gopinathan, S. (2015). Teacher professional development in Singapore: Depicting the landscape. *Psychology, Society and Education, 7*(3), 311–326.

Bautista, A., Wong, J., & Gopinathan, S. (2015). Teacher professional development in Singapore: Depicting the landscape. *Psychology, Society and Education, 7*(3), 311–326.

Beck, C., & Kosnik, C. (2006). *Innovations in teacher education: A social constructivist approach*. Albany, New York: SUNY Press.

Bentler, P. M. (1990). Comparative fit indexes in structural models. *Psychological Bulletin, 107*(2), 238–246.

Bilgin, A., & Balbag, M. Z. (2018). Personal professional development efforts of science and technology teachers in their fields. *Journal of Education in Science, Environment and Health (JESEH), 4*(1), 19–31. DOI:10.21891/jeseh.387477.

Binnewies, C., & Gromer, M. (2012). Creativity and innovation at work: The role of work characteristics and personal initiative. *Psicothema, 24,* 100–105. Retrieved from http://www.psicothema.com/PDF/3985.pdf

Bong, M., & Skaalvik, E. (2003). Academic self-concept and self-efficacy: How different are they really? *Educational Psychology Review, 15*(1), 1–40. http://dx.doi.org/10.1023/A:1021302408382.

Borasi, R., & Finnigan, K. (2010). Entrepreneurial attitudes and behaviors that can help prepare successful change-agents in education. *New Educator, 6,* 1–29. doi:10.1080/1547688X.2010.10399586.

Bourgonjon, J., de Grove, F., de Smet, C., van Looy, J., Soetaert, R., & Valcke, M. (2013). Acceptance of game-based learning by secondary school teachers. *Computers & Education, 67,* 21–35. doi:10.1016/j.compedu.2013.02.010.

Brekelmans, M., Wubbels, T., & van Tartwijk, J. (2005). Teacher–student relationships across the teaching career, chapter 4. *International Journal of Educational Research, 43*(1–2), 55–71.

Brewer, D., & Tierney, W. (2012). Barriers to innovation in the US education. In B. Wildavsky, A. Kelly, & K. Carey (Eds.), *Reinventing higher education: The promise of innovation* (pp. 11–40). Cambridge, MA: Harvard Education Press.

Campbell, C. (2019). *Perspectives and evidence on effective CPD from Canada*. In Shutt, C., & Harrison, S. (Eds), Teacher CPD: International trends, opportunities and challenges (pp. 68–74). London: Chartered College of Teaching. Retrieved from https://chartered.college/international-teacher-cpd-report/

Carmeli, A., Meitar, R., & Weisberg, J. (2006). Self-leadership skills and innovative behavior at work. *International Journal of Manpower, 27,* 75–90. doi:10.1108/01437720610652853.

Chaudhuri, A. R., McCormick, B. D., & Lewis, R., Jr. (2019). Standards-Based Science Institutes: Effective Professional Development That Meets Teacher and District Needs. *Science Educator*, 27(1), 15–23.

Chai, C. S. (2019). Teacher professional development for Science, Technology, Engineering and Mathematics (STEM) education: A review from the perspectives of Technological Pedagogical Content (TPACK). *Asia-Pacific Education Researcher (Springer Science & Business Media B.V.)*, 28(1), 5–13. https://doi-org.ezproxy.vdu.lt:2443/10.1007/s40299-018-0400-7.

Chang, J. C., & Yang, Y. L. (2012). The effect of organization's innovational climate on student's creative self-efficacy and innovative behavior. *Business & Entrepreneurship Journal*, 1(1), 75–100.

Chang, Y. CH. (2018). Analyzing the moderating effect of knowledge innovation of tourism and hospitality department teachers on student creative self-efficacy and innovation behaviors by using hierarchical linear modeling. *Cogent Education*, 5, 1–17. https://doi.org/10.1080/2331186X.2018.1535755.

Clark, N. M., & Zimmerman, B. J. (2014). A social cognitive view of self-regulated learning about health. *Health Education & Behavior*, 41(5), 485–491. https://doi.org/10.1177/1090198114547512.

Cohen, D. K., & Ball, D. L. (2007). Educational innovation and the problem of scale. In B. L. Schneider & S.-K. McDonald (Eds.), *Scale-up in education: Ideas in principle* (Vol. 1, pp. 19–36). Plymouth: Rowman & Littlefield.

Comte, A. (1957). *A general view of positivism*. New York, NY: Speller. (Original work published 1848).

Coser, L. A. (1957). Social conflict and the theory of social change. *The British Journal of Sociology*, 8(3), 197. https://doi.org/10.2307/586859.

Curtis M. (2020). Toward understanding secondary teachers' decisions to adopt geospatial technologies: An examination of Everett Rogers' diffusion of innovation framework. *Journal of Geography*, 119(5), 147–158. doi: 10.1080/00221341.2020.1784252.

Darling-Hammond, L., Hyler, M. E., & Gardner, M. (2017). *Effective teacher professional development*. Palo Alto, CA: Learning Policy Institute.

Darling-Hammond, L., Wei, R. C., Andree, A., Richardson, N., & Orphanos, S. (2009). *Professional learning in the learning profession*. Washington, DC: National Staff Development Council.

De Jong, J., & den Hartog, D. (2010). Measuring innovative work behavior. *Creativity and Innovation Management*, 19(1), 23–36. http://dx.doi.org/10.1111/j.1467-8691.2010.00547.x.

Deary, I. J., Strand, S., Smith, P., & Fernandes, C. (2007). Intelligence and educational achievement. *Intelligence, 35*(1), 13–21. https://doi.org/10.1016/j.intell.2006.02.001.

DeCoito, I. (2006). Innovations in science education: Challenging and changing teachers' roles and beliefs. *Canadian Journal of Science, Mathematics, and Technology Education, 6*(4), 339–350.

Desimone, L. M. (2009). Improving impact studies of teachers' professional development: Toward better conceptualizations and measures. *Educational researcher, 38*(3), 181–199.

Donnelly, D., McGarr, O., & O'Reilly, J. (2011). A framework for teachers' integration of ICT into their classroom practice. *Computers & Education, 57*, 1469–1483. doi:10.1016/j.compedu.2011.02.014.

Drucker, P. (1985). *Innovation and entrepreneurship*. New York: Harper and Row.

Druckman, D., & Donohue, W. (2020). Innovations in social science methodologies: An overview. *American Behavioral Scientist, 64*(1), 3–18. https://doi-org.ezproxy.vdu.lt:2443/10.1177/0002764219859623.

Dunlosky, J., Rawson, K. A., Marsh, E. J., Nathan, M. J., & Willingham, D. T. (2013). Improving students' learning with effective learning techniques: Promising directions from cognitive and educational psychology. *Psychological Science in the Public Interest, 14*, 4–58.

Durkheim, E. (1982). *Rules of sociological method*. New York: Simon & Schuster. (Original work published 1895).

Eccles, J. (1983). Expectancies, values, and academic behaviors. In J. T. Spence (Ed.), *Achievement and achievement motives: Psychological and sociological approaches* (pp. 75–146). San Francisco, CA: W. H. Freeman.

Eccles, J. (2009). Who am I and what am I going to do with my life? Personal and collective identities as motivators of action. *Educational Psychologist, 44*(2), 78–89. http://dx.doi.org/10.1080/00461520902832368.

EL-Deghaidy, H., Mansour, N., Aldahmash, A., & Alshamrani, S. (2015). A Framework for Designing Effective Professional Development: Science Teachers' Perspectives in a Context of Reform. *EURASIA Journal of Mathematics, Science & Technology Education, 11*(6), 1579–1601

Ertesvåg, S. (2014). Teachers' collaborative activity in school-wide interventions. *Social Psychology of Education, 17*(4), 565–588. https://doi-org.ezproxy.vdu.lt:2443/10.1007/s11218-014-9262-x.

Eurydice. (2019). *Lithuania: Initial education for teachers working in early childhood and school education*. Retrieved from: https://eacea.ec.europa.eu/national-policies/eurydice/content/initial-education-teachers-working-early-childhood-and-school-education-43_en#_edn1

Eurydice. (2020). *Sweden: Initial education for teachers working in early childhood and school education.* Retrieved from: https://eacea.ec.europa.eu/national-policies/eurydice/content/initial-education-teachers-working-early-childhood-and-school-education-79_en

Farr, J., & Ford, C. (1990). Individual innovation. In M. West & J. Farr (Eds.), *Managing innovation.* London: Sage.

Fenech, R., Baguant, P., & Alpenidze, O. (2021). The impact of dynamic capabilities on teaching strategies in higher education. *Academy of Strategic Management Journal, 20,* 1–13.

Feng, C., Huang, X., & Zhang, L. (2016). A multilevel study of transformational leadership. Dual organizational change and innovative behavior in groups. *Journal of Organizational Change Management, 29*(6), 855–877.

Field, A. (2000). *Discovering statistics using SPSS for Windows.* London, Thousand Oaks, CA, New Delhi: Sage Publications.

Frost, D. (2006). The concept of 'agency' in leadership for learning. *Leading and Managing, 12*(2), 19–28.

Frost, D. (2008). 'Teacher leadership': Values and voice. *School Leadership & Management: Formerly School Organisation, 28*(4), 337–352. doi:10.1080/13632430802292258.

Frost, D. (2010). Teacher leadership and educational innovations. *Interactive.* Retrieved from: http://www.doiserbia.nb.rs/img/doi/0579-6431/2010/0579-64311002201F.pdf

Furtak, E. M., & Kunter, M. (2012). Effects of autonomy-supportive teaching on student learning and motivation. *Journal of Experimental Education, 80*(3), 284–316. doi:10.1080/00220973.2011.573019.

Garner, P. W., Gabitova, N., Gupta, A., & Wood, T. (2018). Innovations in science education: Infusing social emotional principles into early STEM learning. *Cultural Studies of Science Education, 13*(4), 889–903. https://doi.org/10.1007/s11422-017-9826-0.

George, D., & Mallery, M. (2010). *SPSS for Windows step by step: A simple guide and reference, 17.0 update* (10a ed.). Boston: Pearson.

Gibbons, S., & Silva, O. (2011). School quality, child well-being and parents' satisfaction. *Economics of Education Review, 30*(2), 312–331.

Gilson, L. L., & Litchfield, R. C. (2017). Idea collections: A link between creativity and innovation. *Innovation, 19*(1), 80–85. https://doi.org/10.1080/14479338.2016.1270765.

Goddard, R., Yvonne Goddard, Y., Eun Sook Kim, E.S., & Miller, R. (2015). A theoretical and empirical analysis of the roles of instructional leadership. Teacher collaboration. and collective efficacy beliefs in support of student learning. *American Journal of Education, 121, no. 4* (August 2015), 501–530. https://doi.org/10.1086/681925.

Godin B. (2008). Innovation: The History of a Category. Project on the Intellectual History of Innovation Working Paper No. 1, Montréal: Québec.

Gok, T. (2014). Students' achievement, skill and confidence in using stepwise problem-solving strategies. *Eurasia Journal of Mathematics, Science & Technology Education, 10*(6), 617–624.

Gong, Y., Huang, J. C., & Farh, J. L. (2009). Employee learning orientation, transformational leadership, and employee creativity: The mediating role of employee creative self-efficacy. *Academy of Management Journal, 52*(4), 765–778. doi:10.5465/ amj.2009.43670890.

Good, R. G., Wandersee, J. H., & St. Julien, J. (1993). Cautionary notes on the appeal of the new ism (constructivism) in science education. In K. Tobin (Ed.), *The practice of constructivism in science education* (pp. 71–87). Hillsdale, NJ: Erlbaum.

Grgurović, M. (2014). An application of the diffusion of innovations theory to the investigation of blended language learning. *Innovation in Language Learning and Teaching, 8*(2), 155–170, DOI: 10.1080/17501229.2013.789031.

Gumusluoglu, L., & Ilsev, A. (2009a). Transformational leadership and organizational innovation: The roles of internal and external support for innovation. *Journal of Product Innovation Management, 26,* 264–277. CrossRef | Google Scholar

Gumusluoglu, L., & Ilsev, A. (2009b). Transformational leadership, creativity, and organizational innovation. *Journal of Business Research, 62*(4), 461–473. https://doi.org/10.1016/j.jbusres.2007.07.032. CrossRef | Google Scholar

Halász, G. (2018). Measuring innovation in education: The outcomes of a national education sector innovation survey. *European Journal of Education,* 1–17, https://doi.org/10.1111/ejed.12299.

Hargreaves, A. (2005). Educational change takes ages: Life, career and generational factors in teachers' emotional responses to educational change. *Teaching and Teacher Education, 21,* 967–983.

Hargreaves, A. (2019). Teacher collaboration: 30 years of research on its nature, forms, limitations and effects. *Teachers and Teaching, 25*(5), 603–621. https://doi.org/10.1080/13540602.2019.1639499

Harte, W., & Reitano, P. (2015). Pre-service geography teachers' confidence in geographical subject matter knowledge and teaching geographical skills. *International Research in Geographical and Environmental Education, 24*(3), 223–236, DOI: 10.1080/10382046.2015.1034458.

Hekkert, M. P., Suurs, R., Negro, S. O., Smits, R. E. H. M., & Kuhlmann, S. (2007). Functions of innovation systems: A new approach for analysing technological change. *Technological Forecasting and Social Change, 74*, 413–432. https://doi.org/10.1016/j.techfore.2006.03.002.

Hooper, M. (2016). Developing the TIMSS 2015 context questionnaires. In M. O. Martin, I. V. S. Mullis, & M. Hooper (Eds.), *Methods and procedures in TIMSS 2015* (pp. 2.1–2.8). Retrieved from Boston College, TIMSS & PIRLS International Study Center website: http://timss.bc.edu/publications/timss/2015-methods/chapter-2.html

Horng, J., Hong, J., Chanlin, L., Chang, S., & Chu, H. (2005). Creative teachers and creative teaching strategies. *International Journal of Consumer Studies, 29*, 352–358. doi:10.1111/j.1470-6431.2005.00445.x.

Hosseini, S., & Shirazi, Z. R. H. (2021). Towards teacher innovative work behavior: A conceptual model. *Cogent Education, 8*(1), 1869364. doi:10.1080/2331186X.2020.1869364.

Howell, J. M., Shea, Ch., & Higgins, Ch. A. (2005). Champions of product innovation: Defining, development and validating measurement of champion behaviour. *Journal of Business Venturing, 20*(5), 641–66.

Hsiao, H., Chang, J., Tu, Y., & Chen, S. (2011). The impact of self-efficacy on innovative work behavior for teachers. *International Journal of Social Science and Humanity*, , 31–36. doi:10.7763/IJSSH.2011.V1.6.

Jakubavičius, A., Strazdas, R., & Gečas, K. (2003). *Inovacijos. Procesai, valdymo modeliai, galimybės*. Vilnius: Lietuvos inovacijų centras.

Janiūnaitė, B. (2004). *Edukacinės novacijos ir jų diegimas*. Kaunas: Technologija.

Jansen, M., Scherer, R., & Schroeders, U. (2015). Students' self-concept and self-efficacy in the sciences: Differential relations to antecedents and educational outcomes. *Contemporary Educational Psychology, 41*, 13–24. doi: http://dx.doi.org/10.1016/j.cedpsych.2014.11.002.

Jansen, M., Schroeders, U., & Ludtke, O. (2014). Academic self-concept in science: Multidimensionality, relations to achievement measures, and gender differences. *Learning and Individual Differences, 30*, 11–21. doi: 10.1016/j.lindif.2013.12.003.

Janssen, O. (2000). Job demands, perceptions of effort-reward fairness and innovative work behaviour. *Journal of Occupational and Organizational Psychology, 73*, 287–302.

Janssen, O. (2003). Innovative behavior and job involvement at the price of conflict and less satisfactory relations with co-workers. *Journal of Occupational and Organizational Psychology, 76*, 347–364. doi:10.1348/096317903769647210.

Janssen, O. (2004). How fairness perceptions make innovative behavior more or less stressful. *Journal of Organizational Behavior, 25*(2), 201–215. https://doi.org/10.1002/job.238.

Jung, D., Chow, C., & Wu, A. (2003). The role of transformational leadership in enhancing organizational innovation: Hypotheses and some preliminary findings. *The Leadership Quarterly, 14*(4–5), 525–544. https://doi.org/10.1016/s1048-9843(03)00050-x. CrossRef | Google Scholar

Jung, D., Chow, C., & Wu, A. (2008). Towards understanding the direct and indirect effects of CEOs' transformational leadership on firm innovation. *The Leadership Quarterly, 19*(5), 582–594. CrossRef | Google Scholar

Jurksiene, L., & Pundziene, A. (2016). The relationship between DCs and firm competitive advantage: The mediating role of organizational ambidexterity. *European Business Review, 28*(4), 431–448.

Kaleem, U. I., Waseer, W., A., Hussain, M., & Javed, A. (2018). Re-shaping professional development of university teachers: An analytical study focusing organizational and task factors. *Imperial Journal of Interdisciplinary Research (IJIR), 4*, 237–253.

Kanter, R. (1988). *When a thousand flowers bloom structural, collective, and social conditions for innovation in organizations*. In B. M. Staw & L. L. Cummings (Eds.), *Research in organizational behavior* (Vol. 10, pp. 169–211). Greenwich, CT: JAI Press.

Kareem, M. A., & Alameer, A. A. A. (2019). The impact of dynamic capabilities on organizational effectiveness, management and marketing. *Challenges for the Knowledge Society, 14*(4), 402–418.

Karlberg, M., & Bezzina, C. (2020). The professional development needs of beginning and experienced teachers in four municipalities in Sweden. *Professional Development in Education*. doi: 10.1080/19415257.2020.1712451.

Kaya, V. H., & Elster, D. (2019). Environmental science, technology, engineering, and mathematics pedagogical content knowledge: Teachers' professional development as environmental science, technology, engineering, and mathematics literate individuals in the light of experts' opinions. *Science Education International, 30*(1), 11–20.

Kaya, V. H., & Gödek, Y. (2016). Perspectives in regard to factors affecting the professional development of science teachers. *Journal of Human Sciences, 13*(2), 2625–2640.

Kelly, D. L., Centurino, V., Martin, M. O., & Mullis, I. V. S. (Eds.) (2020). *TIMSS 2019 encyclopedia: Education policy and curriculum in mathematics and science*. Retrieved from: https://timssandpirls.bc.edu/timss2019/encyclopedia/

Kenny, J. (2010). Preparing pre-service primary teachers to teach primary science: A partnership-based approach. *International Journal of Science Education, 32* (10), 1267–1288.

Khoo, K. K., Yeap, J. A. L., & Ramayah, T. (2014). Knowledge absorptive capacity and process innovation: The moderating effect of environmental dynamism. *Journal of Technology Management and Business, 1*(1), 1–20.

Kim, J. Y., Choi, D. S., Sung, C.-S., Joo, Y., & Park, J. Y. (2018). The role of problem solving ability on innovative behavior and opportunity recognition in university students. *Journal of Open Innovation: Technology, Market, and Complexity* 4:4. Retrieved from: https://link.springer.com/article/10.1186/s40852-018-0085-4

Klaeijsen, A., Vermeulen, M., & Martens, R. (2018). Teachers' innovative behaviour: The importance of basic psychological need satisfaction, intrinsic motivation, and occupational self-efficacy. *Scandinavian Journal of Educational Research, 62*(5), 769–782. https://doi-org.ezproxy.vdu.lt:2443/10.1080/00313831.2017.1306803

Kleysen, R., & Street, C. (2001). Towards a multi-dimensional measure of individual innovative behavior. *Journal of Intellectual Capital, 2*(3), 284–296. https://doi.org/10.1108/EUM0000000005660.

Knapp, M. S. (2003). Professional development as policy pathway. *Review of Research in Education, 27*(1), 109–157.

Kolleck, N. (2014). Innovations through networks: Understanding the role of social relations for educational innovations. *Zeitschrift Für Erziehungswissenschaft, 17*(S5), 47–64. https://doi.org/10.1007/s11618-014-0547-9.

Konermann, J. (2011). *Teachers' work engagement: A deeper understanding of the role of job and personal resources in relationship to work engagement, its antecedents, and its outcomes* (Doctoral dissertation). Retrieved from http://doc.utwente.nl/79200/1/thesis_J_Konermann.pdf

Lawlor, J., Marshall, K., & Tangney, B. (2015). Bridge21—exploring the potential to foster intrinsic student motivation through a team-based, technology-mediated learning model. *Technology, Pedagogy and Education, 25*(2). Retrieved January 5, 2017, from http://dx.doi.org/10.1080/1475939X.2015.1023828.

Layton, D. (1986). *Innovators' dilemmas: Recontextualising science and technology education*. In D. Layton (Ed.), *Innovations in science and technology education* (pp. 16–17). UNESCO.

Lemke, J. L. (2001). Articulating communities: Sociocultural perspectives on science education. *Journal of Research in Science Teaching, 38*, 296–316.

Levitt, T. (2002). *Creativity is not enough*. Available from: https://hbr.org/2002/08/creativity-is-not-enough

Liu, X., Gong, X., Wang, F., Sun, R., Gao, Y., Zhang, Y., Zhou, J., & Deng, X. (2017). A new framework of science and technology innovation education for K-12 in Qingdao. In *Proceedings of the 2017 American Society for Engineering Education International Forum ASEE 2017*, Columbus, OH, USA, 28 June 2017.

Loogma, K., Kruusvall, J., & Ümarik, M. (2012). E-learning as innovation: Exploring innovativeness of the VET teachers' community in Estonia. *Computers & Education, 58*, 808–817. doi:10.1016/j.compedu.2011.10.005.

Lownsbrough, S. (2020). Exploring approaches to ensure more effective professional development for English teachers. *RaPAL Journal*, 101, 30–37.

Makri, M., & Scandura, T. A. (2010). Exploring the effects of creative CEO leadership on innovation in high-technology firms. *The Leadership Quarterly, 21*(1), 75–88. https://doi.org/10.1016/j.leaqua.2009.10.006. CrossRef | Google Scholar

Malandrakis, G. (2018). Influencing Greek pre-service teachers' efficacy beliefs and self-confidence to implement the new 'Studies for the Environment' curricula. *Environmental Education Research, 24*(4), 537–563.

Marth, M., Bogner, F. X., & Sotiriou, S. (2018). Professional development in science summer schools: How science motivation and technology interest link in with innovative educational pathways. *International Journal of Learning, Teaching and Educational Research, 17*(5), 47–63. doi:10.26803/ijlter.17.5.4.

Mattar, J. A. (2010). *Constructivism and connectivism in education technology: Active, situated, authentic, experiential, and anchored learning*. Retrieved from https://pdfs.semanticscholar.org/d2c3/a33d7c7fcc5822ff04494863490045f8758e.pdf

McKay, S. R., Millay, L., Allison, E., Byerssmall, E., Wittmann, M. C., Flores, M., Fratini, J., Kumpa, B., Lambert, C., Pandiscio, E. A., & Smith, M. K. (2018). Investing in teachers' leadership capacity: A model from STEM education. *Maine Policy Review, 27*(1), 54–63.

Megan Kolby Noble. (2016). Science teacher confidence. *Science Journal of Education, 4*(1), 9–13. doi: 10.11648/j.sjedu.20160401.12.

Messman, G., & Mulder, R. (2011). Innovative work behaviour in vocational colleges: Understanding how and why innovations are developed. *Vocations and Learning, 4*, 63–84.

Messmann, G., & Mulder, R. H. (2011). Innovative work behaviour in vocational colleges: Understanding how and why innovations are developed. *Vocations and Learning, 4*, 63–84. doi:10.1007/s12186-010-9049-y.

Messmann, G., & Mulder, R. H. (2012). Development of a measurement instrument for innovative work behaviour as a dynamic and context-bound construct. *Human Resource Development International, 15*(1), 43–59. doi:10.1080/13678868.2011.646894.

Messmann, G., & Mulder, R. H. (2014). Exploring the role of target specificity in the facilitation of vocational teachers' innovative work behaviour. *Journal of Occupational & Organizational Psychology, 87*(1), 80–101. https://doi-org.ezproxy.vdu.lt:2443/10.1111/joop.12035

Meyer, J. P., & Allen, N. J. (1997). *Commitment in the workplace: Theory, research, and application*. Thousand Oaks, CA: SAGE.

Meyer, J., Fleckensteina, J., & Köller, O. (2019). Expectancy value interactions and academic achievement: Differential relationships with achievement measures. *Contemporary Educational Psychology, 58*, 58–74.

Meyer, J. P., Stanley, D. J., Herscovitch, L., & Topolnytsky, L. (2002). Affective, continuance, and normative commitment to the organization: A meta-analysis of antecedents, correlates, and consequences. *Journal of Vocational Behavior, 61*, 20–52.

Ming, L., & Ying, Z. Z. (2010). How does organizational commitment affect organizational innovation. In *2010 International Conference on E-Business and E-Government*, Guangzhou, pp. 1164–1170. doi: 10.1109/ICEE.2010.300.

Mitchell, R. M., &Tarter, C. J. (2016). A path analysis of the effects of principal professional orientation towards leadership. *Professional Teacher Behavior, and School Academic Optimism on School Reading Achievement, Societies* 2016, 6(1). 5. https://doi.org/10.3390/soc6010005.

Mokhber, M., bin Wan Ismail, W. K., & Vakilbashi, A. (2015). Effect of transformational leadership and its components on organizational innovation. *Iranian Journal of Management Studies, 8*(2), 221–241.

Mokhber, M., Khairuzzaman, W., & Vakilbashi, A. (2018). Leadership and innovation: The moderator role of organization support for innovative behaviors. *Journal of Management & Organization 24*(1), 108–128.

Mowday, R. T., Porter, L. W., & Steers, R. M. (1982). *Employee organization linkages: The psychology of commitment, absenteeism and turnover*. New York, NY: Academic Press.

Mullis, I. V. S., Martin, M. O., Foy, P., & Hooper, M. (2016). *TIMSS 2015 international results in mathematics*. Retrieved from http://timssandpirls.bc.edu/timss2015/international-results/

Mullis, I. V. S., Martin, M. O., Goh, S., & Cotter, K. (Eds.) (2016). *TIMSS 2015 encyclopedia: education policy and curriculum in mathematics and science*. Retrieved from Boston College, TIMSS & PIRLS International Study Center website: http://timssandpirls.bc.edu/timss2015/encyclopedia/

Mushayikwa, E., & Lubben, F. (2009). Self-directed professional development: Hope for teachers working in deprived environments? *Teaching and Teacher Education, 25*, 375–382. doi:10.1016/j.tate.2008.12.003.

Nagengast, B., Marsh, H. W., Scalas, L. F., Xu, M. K., Hau, K. T., & Trautwein, U. (2011). Who took the "×" out of expectancy-value theory? A psychological mystery, a substantive-methodological synergy, and a cross-national generalization. *Psychological Science, 22*(8), 1058–1066.

Nakata, Y. (2011). Teachers' readiness for promoting learner autonomy: A study of Japanese EFL high school teachers. *Teaching and Teacher Education, 27*, 900–910. doi:10.1016/j.tate.2011.03.001

Naslund, J. A., Aschbrenner, K. A., McHugo, G. J., Unützer, J., Marsch, L. A., & Bartels, S. J. (2019). Exploring opportunities to support mental health care using social media: A survey of social media users with mental illness. *Early Intervention in Psychiatry, 13*(3), 405–413. https://doi-org.ezproxy.vdu.lt:2443/10.1111/eip.12496

National Institute of Education. (2021). Undergraduate programmes. Retrieved from: https://www.nie.edu.sg/teacher-education/undergraduate-programmes/

National Research Council (NRC). (1996). *National science education standards.* Washington, DC: National Academy Press.

National Research Council (NRC). (2000). *Inquiry and the national science education standards: A guide for teaching and learning.* Washington, DC: National Academy Press.

National Research Council. (2012). *A framework for K-12 science education: Practices, crosscutting concepts, and core ideas.* Washington, DC: The National Academies Press.

Ng, T. W. H., & Lucianetti, L. (2016). Goal striving, idiosyncratic deals and job behavior. *Journal of Organizational Behavior, 37*, 41–60. https://doi.org/10.1002/job.2023.

Noefer, K., Stegmaier, R., Molter, B., & Sonntag, K. (2009). Great many things to do and not a minute to spare: Can feedback from supervisors moderate the relationship between skill variety, time pressure, and employees' innovative behavior? *Creativity Research Journal, 21*, 384–393. doi:10.1080/10400410903297964.

O'Sullivan, D. & Dooley, L. (2009). *Applying innovation.* Thousand Oaks: Sage.

OECD (2014). *Measuring innovation in education: A new perspective.* Paris: OECD Publishing.

OECD (2019a). *Measuring innovation in education: A new perspective.* . Paris/Eurostat, Luxembourg: OECD Publishing. https://doi.org/10.1787/9789264304604-en.

OECD. (2019b). *TALIS 2018 results (Volume I): Teachers and school leaders as lifelong learners*. Paris: TALIS, OECD Publishing. Retrieved from: https://www.oecd-ilibrary.org/sites/dd6dd4bc-en/index.html?itemId=/content/component/dd6dd4bc-en

OECD/Eurostat. (2018). *Oslo manual 2018: Guidelines for collecting, reporting and using data on innovation, 4th edition, the measurement of scientific, technological and innovation activities*. Paris/Eurostat, Luxembourg: OECD Publishing. https://doi.org/10.1787/9789264304604-en.

Okada, A. (2013). Scientific literacy in the digital age: Tools, environments and resources for co-inquiry. *European Scientific Journal, 4*, 263–274.

Oke, A., Munshi, N., & Walumbwa, F. (2009). The influence of leadership on innovation processes and activities. *Organizational Dynamics, 38*(1), 64–72. https://doi.org/10.1016/j.orgdyn.2008.10.005.

Osborne, R. J., & Wittrock, M. C. (1983). Learning science: A generative process. *Science Education, 67*, 489–508.

Osunkwo, J.-P. N., & Enyaosah, U. A. (2016). Research and innovations in sciences/science education. *International Journal of Academia, 2*(1), 1–11.

Pajares, F. (2002). *Overview of social cognitive theory and of self-efficacy*. Retrieved January 20, 2006, from http://www.emory.edu/EDUCATION/mfp/eff.html.

Palmer, D. (2005). A motivational view of constructivist-informed teaching. *International Journal of Science Education, 27*(15, 16), 1853–1881.

Pell, T., Galton, M., Steward, S., Page, C., & Hargreaves, L. (2007). Promoting group work at Key Stage 3: Solving an attitudinal crisis among young adolescents? *Research Papers in Education, 22*, 309–332.

Piaget, J. (1978). *The development of thought: Equilibration of cognitive structures*. Oxford, England: Blackwell.

Prasad, B., & Junni, P. (2016). CEO transformational and transactional leadership and organizational innovation: The moderating role of environmental dynamism. *Management Decision, 54*(7), 1542–1568.CrossRef | Google Scholar

Pudjiarti, E. S., & Hutomo, P. T. P. (2020). Innovative work behaviour: An integrative investigation of person-job fit, person-organization fit, and person-group fit. *Business: Theory and Practice, 21*(1), 39–47. https://doi.org/10.3846/btp.2020.9487.

Pudjiarti, E. S., & Hutomo, P. T. P. (2020). Innovative work behaviour: An integrative investigation of person-job fit, person-organization fit, and person-group fit. *Business: Theory and Practice, 21*(1), 39–47. https://doi.org/10.3846/btp.2020.9487.

Pyle, E. J. (1995). Motivation, action, and social cognitive development in informal science education venues: Convergence with the classroom (Doctoral dissertation, University of Georgia, Athens).

Reckwitz, A. (2003). Grundelemente einer Theorie sozialer Praktiken/Basic Elements of a Theory of Social Practices. *Zeitschrift für Soziologie, 32*(4), 282–301. https://doi.org/10.1515/zfsoz-2003-0401

Rietveld, T., & Van Hout, R. (1993). *Statistical techniques for the study of language and language behaviour*. Berlin, New York: Mouton de Gruyter.

Rogers, E. M. (1995). *Diffusion of innovations* (4th ed.). New York: Free Press.

Rogers, E. M. (2002). Diffusion of preventive innovations. *Addictive Behaviors, 27*, 989–993.

Rogers, E. M. (2003). *Diffusion of innovations* (5th ed.). New York: Free Press.

Rohde, T. E., & Thompson, L. A. (2007). Predicting academic achievement with cognitive ability. *Intelligence, 35*(1), 83–92. https://doi.org/10.1016/j.intell.2006.05.004.

Runhaar, P. R. (2008). *Promoting teachers' professional development*. Enschede: University of Twente. https://doi.org/10.3990/1.9789036527514

Rusmansyah, Y. L., Isnawati, I. M. & Prahani, B. K. (2019). Innovative chemistry learning model: Improving the critical thinking skill and self-efficacy of pre-service chemistry teachers. *Journal of Technology and Science Education, 9*(1), 59–76.

Russell, M. G., Nataliya, V., & Smorodinskaya, N. V. (2018). Leveraging complexity for ecosystemic innovation. *Technological Forecasting and Social Change, 136*, 114–131.

Sahin, I. (2006). Detailed review of Rogers' diffusion of innovations theory and educational technology-related studies based on Rogers' theory. *The Turkish Online Journal of Educational Technology, 5*, 14–23.

Schleicher, A. (2019). *TALIS 2018: Insights and interpretations*. OECD. Retrieved from: http://www.oecd.org/education/talis/TALIS2018_insights_and_interpretations.pdf

Schreiber, J. B., Nora, A., Stage, F. K., Barlow, E. A., & King, J. (2006). Reporting structural equation modeling and confirmatory factor analysis results: A Review. *The Journal of Educational Research, 99*, 323–337.

Schröder, A., & Krüger, D. (2019). Social innovation as a driver for new educational practices: Modernising, repairing and transforming the education system. *Sustainability, 11*(4), 1070. https://doi.org/10.3390/su11041070.

Schumpeter, J. (1963). *History of economic analysis.* New York: Oxford University Press.

Schussler, D. L., Poole, I. R., Whitlock, T. W., & Evertson, C. M. (2007). Layers and links: Learning to juggle "one more thing" in the classroom. *Teaching and Teacher Education, 23,* 572–585. doi:10.1016/j.tate.2007.01.016.

Scott, S. G., & Burce, R. A. (1994). Determinants of innovative behavior: A path model of individual innovation in the workplace. *Academy of Management Journal, 37*(3), 580–607.

Serdyukov, P. (2017). Innovation in education: What works, what doesn't, and what to do about it? *Journal of Research in Innovative Teaching & Learning, 10*(1), 4–33.

Sheldrake, R. (2016). Confidence as motivational expressions of interest, utility, and other influences: Exploring under-confidence and over-confidence in science students at secondary school. *International Journal of Educational Research, 76,* 50–65. doi: 10.1016/j.ijer.2015.12.001.

Sherry, L. (1997). The boulder valley internet project: Lessons learned. THE (Technological Horizons in Education) *Journal, 25*(2), 68–73.

Shelton, J. (2011). Education innovation: What it is and why we need more of it, EducationWeek, Sputnik post, September 28, available at: http://blogs.edweek.org/edweek/sputnik/2011/09/education_

Shen, J., Benson, J., & Huang, B. (2014). High-performance work systems and teachers' work performance: The mediating role of quality of working life. *Human Resource Management, 53*(5), 817–833.

Shewbridge, C., et al. (2016). *OECD reviews of school resources: Lithuania 2016,* OECD Reviews of School Resources. Paris: OECD Publishing. Retrieved form: https://doi.org/10.1787/9789264252547-en.

Sims, S., & Fletcher-Wood, H. (2021). Identifying the characteristics of effective teacher professional development: a critical review. *School Effectiveness & School Improvement, 32*(1), 47–63. https://doi-org.ezproxy.vdu.lt:2443/10.1080/09243453.2020.1772841

Sirotin, V., & Arhipova, M. (2015). Cooperation and innovation activity: Study of the relationship at the regional level. In B. Galbraith (Ed.), *Proceedings of the European conference on innovation and entrepreneurship.* Issue 10 (pp. 673–682). Genoa: University of Genoa.

Skaalvik, E. M., & Skaalvik, S. (2004). Self-concept and self-efficacy: A test of the internal/external frame of reference model and predictions of subsequent motivation and achievement. *Psychological Reports, 95,* 1187–1202. doi:10.2466/ pr0.95.3f.1187-1202.

Skolverket [Swedish National Agency for Education]. (2021). Apply for certification with a foreign diploma. Retrieved from: https://www.skolverket.se/regler-och-ansvar/lararlegitimation-och-forskollararlegitimation/lararlegitimation-och-forskollararlegitimation-med-utlandsk-examen/lankade-puffar/certification-of-teachers-with-a-foreign-diploma

So, K. (2013). Knowledge construction among teachers within a community based on inquiry as stance. *Teaching and Teacher Education, 29*, 188–196. doi:10.1016/j.tate.2012.10.005.

Smith, R., Ralston, N. C., Naegele, Z., & Waggoner, J. (2020). Team Teaching and Learning: A Model of Effective Professional Development for Teachers. *Professional Educator, 43*(1), 80–90.

Spillane, J. P., Halverson, R., & Diamond, J. B. (2001). Investigating school leadership practice: A distributed perspective. *Educational Researcher, 30*(3), 23–28.

Sun, Y., & Huang, J. (2019). Psychological capital and innovative behavior: Mediating effect of psychological safety. *Social Behavior and Personality: An International Journal, 47*(9), 1–7. doi:10.2224/sbp.8204.

Teerling, A., Bernholt, A., Igler, J., Schlitter, T., Ohle-Peters, A., McElvanyb, N., & Koller, O. (2020). The attitude does matter: The role of principals' and teachers' concerns in an implementation process. *International Journal of Educational Research, 100*(2020), 101528.

Thomas, A., & Thorne, G. (2009). *How to increase higher order thinking.* Metairie, LA: Center for Development and Learning. Retrieved Dec. 7, 2009, from http://www.cdl.org/resource-library/articles/HOT.php?type=subject&id=18

Thurlings, M., Evers, A., & Vermeulen, M. (2015). Towards a model of explaining teachers' innovative behavior: A literature review. *Review of Educational Research, 85*, 430–471. doi:10.3102/0034654314557949.

Timucin, M. (2009). Diffusion of technological innovation in a foreign languages unit in Turkey: A focus on risk-aversive teachers. Technology, *Pedagogy and Education, 18*(1), 75–86. https://doi.org/10.1080/14759390802704121

Towndrow, P. A., Silver, R. E., & Albright, J. (2010). Setting expectations for educational innovations. *Journal of Educational Change, 11*, 425–455. https://doi.org/10.1007/s10833-009-9119-9.

Train, T. L., & Miyamoto, Y. J. (2017). Encouraging science communication in an undergraduate curriculum improves students' perceptions and confidence. *Journal of College Science Teaching, 46*, 76–83.

Trapitsin, S., Granichin, O., Granichina, O., & Zharova, M. (2018). Innovative behavior of teachers: Definition and analysis. In *18th PCSF 2018 professional culture of the specialist of the future. The European proceedings of social & behavioural sciences.*

Trautwein, U., Marsh, H. W., Nagengast, B., Lüdtke, O., Nagy, G., & Jonkmann, K. (2012). Probing for the multiplicative term in modern expectancy–value theory: A latent interaction modeling study. *Journal of Educational Psychology, 104*(3), 763–777.

Tsai, C. T., & Tseng, W. W. (2010). A research agenda of transformational leadership and innovative behavior for the hospitality industry: An integrated multilevel model. Annual international council on hotels restaurants and institutional education conference, Puerto Rico, USA.

Tuominen, T., & Toivonen, M. (2011). Studying innovation and change activities in KIBS through the lens of innovative behaviour. *International Journal of Innovation Management, 15*, 393–422.

Uddin, M. A., Fan, L., & Das, A. K. (2017). A study of the impact of transformational leadership. Organizational learning. and knowledge management on organizational innovation. *Management Dynamics, 16*(2), 42–54.

Ullman, J. B. (2001). Structural equation modeling. In B. G. Tabachnick & L. S. Fidell (Eds.), *Using multivariate statistics* (4th ed.). Needham Heights, MA: Allyn & Bacon.

Valentinavičius, S. (2006). Inovacinio verslo plėtra: Problemos ir galimybės. *Ekonomika, 74*, 108–128.

Van der Heijden, B. I. J. M., Gorgievski, M. J., & De Lange, A. H. (2015). Learning at the workplace and sustainable employability: A multi-source model moderated by age. *European Journal of Work and Organizational Psychology.* doi: 10.1080/1359432X.2015.1007130.

Viljaranta, J., Tolvanen, A., Aunola, K., & Nurmi, J.-E. (2014). The developmental dynamics between interest, self-concept of ability, and academic performance. *Scandinavian Journal of Educational Research, 58*(6), 734–756. http://dx.doi.org/10.1080/00313831.2014.904419.

Vincent-Lancrin, S., et al. (2019). *Measuring innovation in education 2019: What has changed in the classroom?*, Educational research and innovation. Paris: OECD Publishing. https://doi.org/10.1787/9789264311671-en.

Vygotsky, Lev S. (1978). *Mind in society: The development of higher mental processes.* Cambridge, MA: Harvard University Press.

Wallin, A., Kettunen, P., Johansson, P. M., Jonsdottir, I. H., Nilsson, C., Nilsson, M., ... Georg Kuhn, H. (2018). Cognitive medicine—a new approach in health care science. *BMC Psychiatry, 18*(1), 0–5. https://doi.org/10.1186/s12888-018-1615-0.

Weiss, I. R., & Pasley, J. D. (2006). *Scaling up instructional improvement through teacher professional development: Insights from the local systemic change initiative*. Philadelphia, PA: Consortium for Policy Research in Education (CPRE) Policy Briefs.

Welbourne, T. M., Johnson, D. E., & Erez, A. (1998). The role-based performance scale: Validity analysis of a theory-based measure. *Academy of Management Journal, 41*, 540–555. http://dx.doi.org/10.2307/256941.

West, M. A., & Farr, J. L. (1990). Innovation at work. In M. A. West & J. L. Farr (Eds.), *Innovation and creativity at work* (pp. 03–13). New York: John Wiley and Sons.

Widmann, A., & Mulder, R. (2018). Team learning behaviours and innovative work behaviour in work teams. *European Journal of Innovation Management, 21*(3), 501–520. https://doi.org/10.1108/EJIM-12-2017-0194.

Wigfield, A., & Eccles, J. S. (2000). Expectancy-value theory of motivation. *Contemporary Educational Psychology, 25*, 68–81.

Yang, S. C., & Huang, Y.-F. (2008). A study of high school English teachers' behavior, concerns and beliefs in integrating information technology into English instruction. *Computers in Human Behavior, 24*, 1085–1103. doi:10.1016/j.chb.2007.03.009.

Yates, S. (1990). How confident are primary school teachers in teaching science? *Research in Science Education, 20*, 300–305.

Yürük, N. (2011). The predictors of preservice elementary teachers' anxiety about teaching science. *Journal of Baltic Science Education, 10*(1), 17–26.

Zell, E., & Krizan, Z. (2014). Do people have insight into their abilities? A metasynthesis. *Perspectives on Psychological Science, 9*(2), 111–125. http://dx.doi.org/10.1177/1745691613518075.

Zhou, J., & George, J. M. (2001). When job dissatisfaction leads to creativity: Encouraging the expression of voice. *Academy of Management Journal, 44*(4), 682–696. https://doi.org/10.2307/3069410.

Zulu, F.-Q. B., & Bertram, C. (2019). Collaboration and teacher knowledge in a teacher learning community: A case of mathematics teachers in Kwazulu-Natal province. *South African Journal of Higher Education, 33*(5), 112–129. https://doi-org.ezproxy.vdu.lt:2443/10.20853/33-5-3595.

**ERZIEHUNGSKONZEPTIONEN UND PRAXIS**
Herausgeben von Gerd-Bodo von Carlsburg

| | |
|---|---|
| Band 1 | Barbara Hellinge / Manfred Jourdan / Hubertus Maier-Hein: Kleine Pädagogik der Antike. 1984. |
| Band 2 | Siegfried Prell: Handlungsorientierte Schulbegleitforschung. Anleitung, Durchführung und Evaluation. 1984. |
| Band 3 | Gerd-Bodo Reinert: Leitbild Gesamtschule versus Gymnasium? Eine Problemskizze. 1984. |
| Band 4 | Ingeborg Wagner: Aufmerksamkeitsförderung im Unterricht. Hilfen durch Lehrertraining. 1984. |
| Band 5 | Peter Struck: Pädagogische Bindungen. Zur Optimierung von Lehrerverhalten im Schulalltag. 1984. |
| Band 6 | Wolfgang Sehringer (Hrsg.): Lernwelten und Instruktionsformen. 1986. |
| Band 7 | Gerd-Bodo Reinert (Hrsg.): Kindgemäße Erziehung. 1986. |
| Band 8 | Heinrich Walther: Testament eines Schulleiters. 1986. |
| Band 9 | Gerd-Bodo Reinert / Rainer Dieterich (Hrsg.): Theorie und Wirklichkeit – Studien zum Lehrerhandeln zwischen Unterrichtstheorie und Alltagsroutine. 1987. |
| Band 10 | Jörg Petersen / Gerhard Priesemann: Einführung in die Unterrichtswissenschaft. Teil 1: Sprache und Anschauung. 2., überarb. Aufl. 1992. |
| Band 11 | Jörg Petersen / Gerhard Priesemann: Einführung in die Unterrichtswissenschaft. Teil 2: Handlung und Erkenntnis. 1992. |
| Band 12 | Wolfgang Hammer: Schulverwaltung im Spannungsfeld von Pädagogik und Gesellschaft. 1988. |
| Band 13 | Werner Jünger: Schulunlust. Messung – Genese – Intervention. 1988. |
| Band 14 | Jörg Petersen / Gerhard Priesemann: Unterricht als regelgeleiteter Handlungszusammenhang. Ein Beitrag zur Verständigung über Unterricht. 1988. |
| Band 15 | Wolf-Dieter Hasenclever (Hrsg.): Pädagogik und Psychoanalyse. Marienauer Symposion zum 100. Geburtstag Gertrud Bondys. 1990. |
| Band 16 | Jörg Petersen / Gerd-Bodo Reinert / Erwin Stephan: Betrifft: Hausaufgaben. Ein Überblick über die didaktische Diskussion für Elternhaus und Schule. 1990. |
| Band 17 | Rudolf G. Büttner / Gerd-Bodo Reinert (Hrsg.): Schule und Identität im Wandel. Biographien und Begebenheiten aus dem Schulalltag zum Thema Identitätsentwicklung. 1991. |
| Band 18 | Eva Maria Waibel: Von der Suchtprävention zur Gesundheitsförderung in der Schule. Der lange Weg der kleinen Schritte. 3. Aufl. 1994. |
| Band 19 | Heike Biermann: Chancengerechtigkeit in der Grundschule – Anspruch und Wirklichkeit. 1992. |
| Band 20 | Wolf-Dieter Hasenclever (Hrsg.): Reformpädagogik heute: Wege der Erziehung zum ökologischen Humanismus. 2. Marienauer Symposion zum 100. Geburtstag von Max Bondy. 1993. 2., durchges. Aufl. 1998. |

Band 21 Bernd Arnold: Medienerziehung und moralische Entwicklung von Kindern. Eine medienpädagogische Untersuchung zur Moral im Fernsehen am Beispiel einer Serie für Kinder im Umfeld der Werbung. 1993.

Band 22 Dimitrios Chatzidimou: Hausaufgaben konkret. Eine empirische Untersuchung an deutschen und griechischen Schulen der Sekundarstufen. 1994.

Band 23 Klaus Knauer: Diagnostik im pädagogischen Prozeß. Eine didaktisch-diagnostische Handreichung für den Fachlehrer. 1994.

Band 24 Jörg Petersen / Gerd-Bodo Reinert (Hrsg.): Lehren und Lernen im Umfeld neuer Technologien. Reflexionen vor Ort. 1994.

Band 25 Stefanie Voigt: Biologisch-pädagogisches Denken in der Theorie. 1994.

Band 26 Stefanie Voigt: Biologisch-pädagogisches Denken in der Praxis. 1994.

Band 27 Reinhard Fatke / Horst Scarbath: Pioniere Psychoanalytischer Pädagogik. 1995.

Band 28 Rudolf G. Büttner / Gerd-Bodo Reinert (Hrsg.): Naturschutz in Theorie und Praxis. Mit Beispielen zum Tier-, Landschafts- und Gewässerschutz. 1995.

Band 29 Dimitrios Chatzidimou / Eleni Taratori: Hausaufgaben. Einstellungen deutscher und griechischer Lehrer. 1995.

Band 30 Bernd Weyh: Vernunft und Verstehen: Hans-Georg Gadamers anthropologische Hermeneutikkonzeption. 1995.

Band 31 Helmut Arndt / Henner Müller-Holtz (Hrsg.): Schulerfahrungen – Lebenserfahrungen. Anspruch und Wirklichkeit von Bildung und Erziehung heute. Reformpädagogik auf dem Prüfstand. 2. Aufl. 1996.

Band 32 Karlheinz Biller: Bildung erwerben in Unterricht, Schule und Familie. Begründung – Bausteine – Beispiele. 1996.

Band 33 Ruth Allgäuer: Evaluation macht uns stark! Zur Unverzichtbarkeit von Praxisforschung im schulischen Alltag. 1997. 2., durchges. Aufl. 1998.

Band 34 Christel Senges: Das Symbol des Drachen als Ausdruck einer Konfliktgestaltung in der Sandspieltherapie. Ergebnisse aus einer Praxis für analytische Psychotherapie von Kindern und Jugendlichen. 1998.

Band 35 Achim Dehnert: Untersuchung der Selbstmodelle von Managern. 1997.

Band 36 Shen-Keng Yang: Comparison, Understanding and Teacher Education in International Perspective. Edited and introduced by Gerhard W. Schnaitmann. 1998.

Band 37 Johann Amos Comenius: Allverbesserung (Panorthosia). Eingeleitet, übersetzt und erläutert von Franz Hofmann. 1998.

Band 38 Edeltrud Ditter-Stolz: Zeitgenössische Musik nach 1945 im Musikunterricht der Sekundarstufe I. 1999.

Band 39 Manfred Luketic: Elektrotechnische Lernsoftware für den Technikunterricht an Hauptschulen. 1999.

Band 40 Gerhard Baltes / Brigitta Eckert: Differente Bildungsorte in systemischer Vernetzung. Eine Antwort auf das Problem der funktionellen Differenzierung in der Kooperation zwischen Jugendarbeit und Schule. 1999.

Band 41 Roswit Strittmatter: Soziales Lernen. Ein Förderkonzept für sehbehinderte Schüler. 1999.

Band 42 Thomas H. Häcker: Widerstände in Lehr-Lern-Prozessen. Eine explorative Studie zur pädagogischen Weiterbildung von Lehrkräften. 1999.

Band 43  Sabine Andresen / Bärbel Schön (Hrsg.): Lehrerbildung für morgen. Wissenschaftlicher Nachwuchs stellt sich vor. 1999.

Band 44  Ernst Begemann: Lernen verstehen – Verstehen lernen. Zeitgemäße Einsichten für Lehrer und Eltern. Mit Beiträgen von Heinrich Bauersfeld. 2000.

Band 45  Günter Ramachers: Das intrapersonale Todeskonzept als Teil sozialer Wirklichkeit. 2000.

Band 46  Christoph Dönges: Lebensweltliche Erfahrung statt empirischer Enteignung. Grenzen und Alternativen empirischer Konzepte in der (Sonder-)Pädagogik. 2000.

Band 47  Michael Luley: Eine kleine Geschichte des deutschen Schulbaus. Vom späten 18. Jahrhundert bis zur Gegenwart. 2000.

Band 48  Helmut Arndt / Henner Müller-Holtz (Hrsg.): Herausforderungen an die Pädagogik aufgrund des gegenwärtigen gesellschaftlichen Wandels. Bildung und Erziehung am Beginn des 3. Jahrtausends. 2000.

Band 49  Johann Amos Comenius: Allermahnung (Pannuthesia). Eingeleitet, übersetzt und erläutert von Franz Hofmann. 2001.

Band 50  Hans-Peter Spittler-Massolle: Blindheit und blindenpädagogischer Blick. Der *Brief über die Blinden zum Gebrauch für die Sehenden* von Denis Diderot und seine Bedeutung für den Begriff von Blindheit. 2001.

Band 51  Eva Rass: Kindliches Erleben bei Wahrnehmungsproblemen. Möglichkeiten einer selbst- psychologisch ausgerichteten Pädagogik und Psychotherapie bei sublimen und unerkannten Schwächen in der sensorischen Integration. 2002.

Band 52  Bruno Hamann: Neue Herausforderungen für eine zeitgemäße und zukunftsorientierte Schule. Unter Mitarbeit von Birgitta Hamann. 2002.

Band 53  Johann Amos Comenius: Allerleuchtung (Panaugia). Eingeleitet, übersetzt und erläutert von Franz Hofmann. 2002.

Band 54  Bernd Sixtus: Alasdair MacIntyres Tugendenlehre von *After Virtue* als Beitrag zum Disput über universalistische Erziehungsziele. 2002.

Band 55  Elke Wagner: Sehbehinderung und Soziale Kompetenz. Entwicklung und Erprobung eines Konzeptes. 2003.

Band 56  Jutta Rymarczyk / Helga Haudeck: *In Search of The Active Learner*. Untersuchungen zu Fremdsprachenunterricht, bilingualen und inter- disziplinären Kontexten. 2003.

Band 57  Gerhard W. Schnaitmann: Forschungsmethoden in der Erziehungswissenschaft. Zum Verhältnis von qualitativen und quantitativen Methoden in der Lernforschung an einem Beispiel der Lernstrategienforschung. 2004.

Band 58  Bernd Schwarz / Thomas Eckert (Hrsg.): Erziehung und Bildung nach TIMSS und PISA. 2004.

Band 59  Werner Sacher / Alban Schraut (Hrsg.): Volkserzieher in dürftiger Zeit. Studien über Leben und Wirken Eduard Sprangers. 2004.

Band 60  Dorothee Dahl: Interdisziplinär geprägte Symbolik in der visuellen Kommunikation. Tendenzen therapeutisch-kunstpädagogischer Unterrichts- modelle vor dem Hintergrund multimedialer Zeitstrukturen. 2005.

Band 61  Gerd-Bodo von Carlsburg / Marian Heitger (Hrsg.): Der Lehrer – ein (un)möglicher Beruf. 2005.

| | |
|---|---|
| Band 62 | Bruno Hamann: Pädagogische Anthropologie. Theorien – Modelle – Strukturen. Eine Einführung. 4., überarbeitete und ergänzte Auflage. 2005. |
| Band 63 | Airi Liimets: Bestimmung des lernenden Menschen auf dem Wege der Reflexion über den Lernstil. 2005. |
| Band 64 | Cornelia Matz: Vorbilder in den Medien. Ihre Wirkungen und Folgen für Heranwachsende. 2005. |
| Band 65 | Birgitta Hamann: Grundfragen der Literaturdidaktik und zentrale Aspekte des Deutschunterrichts. 2005. |
| Band 66 | Ralph Olsen / Hans-Bernhard Petermann / Jutta Rymarczyk (Hrsg.): Intertextualität und Bildung – didaktische und fachliche Perspektiven. 2006. |
| Band 67 | Bruno Hamann: Bildungssystem und Lehrerbildung im Fokus aktueller Diskussionen. Bestandsaufnahme und Perspektiven. 2006. |
| Band 68 | Ingeborg Seitz: Heterogenität als Chance. Lehrerprofessionalität im Wandel. 2007. |
| Band 69 | Margret Ruep / Gustav Keller: Schulevaluation. Grundlagen, Methoden, Wirksamkeit. 2007. |
| Band 70 | Harald Schweizer: Krach oder Grammatik? Streitschrift für einen revidierten Sprachunterricht. Kritik und Vorschläge. 2008. |
| Band 71 | Martina Becker / Gerd-Bodo von Carlsburg / Helmut Wehr (Hrsg.): Seelische Gesundheit und gelungenes Leben. Perspektiven der Humanistischen Psychologie und Humanistischen Pädagogik. Ein Handbuch. 2008. |
| Band 72 | Sigvard Clasen: Bildung im Licht von Beschäftigung und Wachstum. Wohin bewegt sich Deutschland? 2009. |
| Band 73 | Gerd-Bodo von Carlsburg: Enkulturation und Bildung. Fundament sozialer Kompetenz. 2009. |
| Band 74 | Hermann-Josef Wilbert: Musikunterricht im Rückblick. Eine alternative Musikdidaktik. 2009. |
| Band 75 | Britta Klopsch: Fremdevaluation im Rahmen der Qualitätsentwicklung und -sicherung. Eine Evaluation der Qualifizierung baden-württembergischer Fremdevaluatorinnen und Fremdevaluatoren. 2009. |
| Band 76 | Leonard Wehr: Partizipatorisches Marketing privater Hochschulen. Corporate Identity als Ziel von Bildungsmarketing. 2011. |
| Band 77 | Konstantinos D. Chatzidimou: Microteaching als erlebnis- und handlungsorientierte Methode im Rahmen der Lehrerausbildung und der Didaktik. Eine theoretische und empirische Untersuchung. 2012. |
| Band 78 | Miriam Lange: Befähigen, befähigt werden, sich befähigen – Eine Auseinandersetzung mit dem Capability Approach. 2014. |
| Band 79 | Eva Rass (Hrsg.): Comenius: Seiner Zeit weit voraus...! Die Entdeckung der Kindheit als grundlegende Entwicklungsphase. 2014. |
| Band 80 | Cornelia Frech-Becker: Disziplin durch Bildung – ein vergessener Zusammenhang. Eine historisch-systematische Untersuchung aus antinomischer Perspektive als Grundlage für ein bildungstheoretisches Verständnis des Disziplinproblems. 2015. |
| Band 81 | Rüdiger Funiok / Harald Schöndorf (Hrsg.): Ignatius von Loyola und die Pädagogik der Jesuiten. Ein Modell für Schule und Persönlichkeitsbildung. 2017. |

Band 82   Lara Rodríguez Sieweke (ed.): Learning Scenarios for Social and Cultural Change. *Bildung* Through Academic Teaching. 2017.

Band 83   Fred Maurer: Leseförderung durch Kriminalliteratur. Deutschdidaktische Annäherungen an ein verkanntes und vernachlässigtes Genre. 2019.

Band 84   Otilia Clipa / Erica Cîmpan: Social-Emotional Competences of Preschoolers. The Impact of Outdoor Educational Activities. 2020.

Band 85   Mariana Sirotová / Veronika Michvocíková /Marián Hosťovecký: Serious Games in University Education of Future Teachers. 2021.

Band 86   Valdonė Indrašienė/ Violeta Jegelevičienė / Odeta Merfeldaitė / Daiva Penkauskienė / Jolanta Pivorienė /Asta Railienė / Justinas Sadauskas / Natalija Valavičienė: Critical Thinking in Higher Education and Labour Market. 2021.

Band 87   Jan Rolf Friederichs: Schulreformen als Antwort auf gesellschaftliche Veränderungen. Die Gemeinschaftsschule im Spannungsfeld der Bildungspolitik in Baden-Württemberg. 2021.

Band 88   Otilia Clipa (ed.): Challenges in Education – Policies, Practice and Research. 2021.

Band 89   Palmira Pečiuliauskienė / Lina Kaminskienė: Science Teachers' Innovative Work Behavior: Factors and Actors. 2022.

www.peterlang.com